IEEE Recommended Practice for the Design of Reliable Industrial and Commercial Power Systems

Published by
The Institute of Electrical and Electronics Engineers, Inc

Distributed in cooperation with
Wiley-Interscience, a division of John Wiley & Sons, Inc

IEEE
Std 493-1980

IEEE Recommended Practice for the Design of Reliable Industrial and Commercial Power Systems

Sponsor
Power System Technologies Committee
of the
IEEE Industry Applications Society

ISBN 0-471-09261-4

Library of Congress Catalog Number 80-83819

Approved December 20, 1979
IEEE Standards Board

Foreword

(This Foreword is not a part of IEEE Std 493-1980, IEEE Recommended Practice for the Design of Reliable Industrial and Commercial Power Systems.)

The design of reliable industrial and commercial power systems is of considerable interest to many people. Prior to 1962 a qualitative viewpoint was taken when attempting to achieve this objective. The need for a quantitative approach was first recognized in the early 1960s when a small group of pioneers led by W.H. Dickinson organized an extensive AIEE survey of the reliability of electric equipment in industrial plants. The survey of AIEE taken in 1962 was followed by several IEEE reliability surveys which were published during 1973 through 1979.

Tutorial reliability sessions on the design of industrial and commercial power systems were conducted at the 1971 and 1976 IEEE Industry and Commercial Power Systems Technical Conferences. A need has existed for some time that the pertinent tutorial reliability material, equipment reliability data, and cost of power outage data be assembled into a single book that could be useful in the design of reliable industrial and commercial power systems.

Comments are invited on this recommended practice and also suggestions for additional material. These comments and suggestions should be addressed to:

Secretary
IEEE Standards Board
The Institute of Electrical and Electronics Engineers, Inc
345 E. 47th St
New York, NY 10017

This recommended practice was prepared by a working group of the Power Systems Reliability Subcommittee, Power Systems Support Committee, Industrial Power Systems Department of the IEEE Industry Application Society.

At the time it recommended these practices the working group of the Power System Technologies Committee had the following members and contributors:

C.R. Heising, *Chairman*

C. E. Becker
P. E. Gannon
D. W. McWilliams
A. D. Patton
C. Singh
S. J. Wells

Other contributors to the nine IEEE Committee Reports on equipment reliability in the Appendixes are:

J. W. Aquilino
M. F. Chamow
W. H. Dickinson
B. G. Douglas
I. Harley
M. D. Harris
D. Kilpatrick
R. T. Kulvicki
P. O'Donnell
R. W. Parisian
W. J. Pearce†
W. L. Stebbins
H.T. Wane

Contents

SECTION PAGE

SECTION PAGE

APPENDIXES

FIGURES

1. Introduction

1.1 Objectives and Scope. The objective of this book is to present the fundamentals of reliability analysis as it applies to the planning and design of industrial and commercial electric power distribution systems. The text material is primarily directed toward consulting plant electrical engineers.

The design of reliable industrial and commercial power distribution systems is important because of the high costs associated with power outages. There is a need to be able to consider the cost of power outages when making design decisions for new power distribution systems and to be able to make quantitative cost versus reliability tradeoff studies. The lack of credible data concerning equipment reliability and the cost of power outages has hindered engineers in making such studies.

The authors of this book have attempted to provide sufficient information so that reliability analyses can be performed on power systems without need for cross references to other texts. Included are:

(1) Basic concepts of reliability analysis by probability methods

(2) Fundamentals of power system reliability evaluation

(3) Economic evaluation of reliability

(4) Cost of power outage data

(5) Equipment reliability data

(6) Examples of reliability analysis

In addition, discussion and information are provided on:

(1) Emergency and standby power

(2) Electrical preventive maintenance

(3) Evaluating and improving reliability of existing plant

A quantitative reliability analysis includes making a disciplined evaluation of alternate power distribution system design choices. When costs of power outages at the various building and plant locations are factored into the evaluation, the decisions can be based upon *total owning cost over the useful life of the equipment* rather than simply *first cost* of the system. The material in this book should enable engineers to make more use of quantitative cost versus reliability tradeoff studies during the design of industrial and commercial power systems.

1.2 IEEE Reliability Surveys of Industrial Plants. From 1973 through 1979 the Power Systems Reliability Subcommittee of the Power Systems Support Committee, Industrial Power Systems Department, IEEE Industry Applications Society, conducted and published the results of extensive surveys of the reliability of electrical equipment in industrial plants and also the cost of power outages for both industrial plants and commercial buildings. The results from these surveys have been published in nine IEEE committee reports. The most important results from these surveys are summarized in Chapters 2, 3 and 5 of this book. In addition, the nine IEEE committee reports are reprinted in Appendixes A, B, C, D, and E.

The IEEE survey reliability data provide historical experience to those who have not been able to collect their own data. Such data can be an aid in analyzing, designing, or redesigning a power distribution system and can provide a basis for quantitative cost comparisons between alternate designs.

1.3 How to Use This Book. This book is primarily directed toward consulting engineers and plant electrical engineers and covers the fundamentals of reliability analysis as it applies to the planning and design of industrial and commercial electric power distribution systems. The methods of reliability analysis are based upon probability and statistics. Some users of this book may wish to read Chapter 8 on basic probability concepts before reading Chapter 2 on planning and design. Other users may wish to start with Chapter 2 and not wish to attempt to fully understand the derivation of the statistical formulas given in 2.1.9, Table 1.

The most important parts of planning and design are covered in 2.1 and 2.2 on fundamentals of power system reliability evaluation and on the economic evaluation of reliability. Chapter 7 gives six examples using these methods of analysis. These examples cover some of the most common decisions that engineers are facing when designing a power distribution system. Some discussion on the limitations of reliability and availability predictions are given in the latter part of 7.1.

Those wishing to obtain equipment reliability data should go to Chapter 3. Those wishing to obtain data on the cost of electrical interruptions to industrial plants or commercial buildings should consult 2.2. Any data on costs may need to be updated to take into account the effects of inflation.

The importance of electrical preventive maintenance in planning and design is covered in 2.3 and 2.4. Chapter 5 discusses the subject in further detail and contains data showing the effect of maintenance quality on equipment failure rates.

Many reliability studies need to be followed up by considerations for emergency and standby power. This subject is covered in Chapter 6 and may also be considered part of planning and design.

An approach to evaluating and upgrading the reliability of an existing plant is presented in Chapter 4. Some users of this book may wish to start with this Chapter.

2. Planning and Design

2.1 Fundamentals of Power System Reliability Evaluation

2.1.1 Reliability Evaluation Fundamentals. Fundamentals necessary for a quantitative reliability evaluation in electric power systems include definitions of basic terms, discussions of useful measures of system reliability and the basic data needed to compute these indexes, and a description of the procedure for system reliability analysis including computation of quantitative reliability indexes.

2.1.2 Power System Design Considerations. An important aspect of power system design involves consideration of the service reliability requirements of loads which are to be supplied and the service reliability which will be provided by any proposed system. System reliability assessment and evaluation methods based on probability theory which allow the reliability of a proposed system to be assessed quantitatively are finding wide application today. Such methods permit consistent, defensible, and unbiased assessments of system reliability which are not otherwise possible.

The quantitative reliability evaluation methods presented here permit reliability indexes for any electric power system to be computed from knowledge of the reliability performance of the constituent components of the system. Thus, alternative system designs can be studied to evaluate the impact on service reliability and cost of changes in component reliability, system configuration, protection and switching scheme, or system operating policy including maintenance practice.

2.1.3 Definitions. The definitions presented here include those used in the survey of the reliability of electric equipment in industrial plants.[1] [1] The definitions presented are not exhaustive, but do provide much of the required nomenclature for discussions of power system reliability.

availability. This term may apply either to the performance of individual components or to that of a system. Availability is defined to be the long-term average

[1] Numbers in brackets refer to the references listed in 2.5.

fraction of time that a component or system is in service satisfactorily performing its intended function. An alternative and equivalent definition for availability is the steady-state probability that a component or system is in service.

component. A piece of equipment, a line or circuit, or a section of a line or circuit, or a group of items which is viewed as an entity for purposes of reliability evaluation.

expected interruption duration. The expected, or average, duration of a single load interruption event.

exposure time. The time during which a component is performing its intended function and is subject to failure.

failure. Any trouble with a power system component that causes any of the following to occur:

(1) Partial or complete plant shutdown, or below-standard plant operation
(2) Unacceptable performance of user's equipment
(3) Operation of the electrical protective relaying or emergency operation of the plant electrical system
(4) Deenergization of any electric circuit or equipment

A failure on a public utility supply system may cause the user to have either of the following:

(1) A power interruption or loss of service
(2) A deviation from normal voltage or frequency of sufficient magnitude or duration

A failure on an in-plant component causes a forced outage of the component, that is, the component is unable to perform its intended function until it is repaired or replaced. The terms *failure* and *forced outage* are often used synonymously.

failure rate (forced outage rate). The mean number of failures of a component per unit exposure time. Usually exposure time is expressed in years and failure rate is given in failures per year.

forced unavailability. The long-term average fraction of time that a component or system is out of service due to failures.

interruption. The loss of electric power supply to one or more loads.

interruption frequency. The expected (average) number of power interruptions to a load per unit time, usually expressed as interruptions per year.

outage. The state of a component or system when it is not available to properly perform its intended function.

repair time. The repair time of a failed component or the duration of a failure is the clock time from the occurrence of the failure to the time when the component is restored to service, either by repair of the failed component or by substitution of a spare component for the failed component. It is not the time required to restore service to a load by putting alternate circuits into operation. It includes time for diagnosing the trouble, locating the failed component, waiting for parts, repairing or replacing, testing, and restoring the component to service. The terms *repair time* and *forced outage duration* are often used synonymously.

scheduled outage. An outage that results when a component is deliberately taken out of service at a selected time, usually for purposes of construction, maintenance, or repair.

scheduled outage duration. The period from the initiation of a scheduled outage until construction, preventive maintenance, or repair work is completed and the affected component is made available to perform its intended function.

scheduled outage rate. The mean number of scheduled outages of a component per unit exposure time.

switching time. The period from the time a switching operation is required due to a component failure until that switching operation is completed. Switching operations include such operations as: throwover to an alternate circuit, opening or closing a sectionalizing switch or circuit breaker, reclosing a circuit breaker following a trip out due to a temporary fault, etc.

system. A group of components connected or associated in a fixed configuration to perform a specified function of distributing power.

unavailability. The long-term average fraction of time that a component or system is out of service due to failures or scheduled outages. An alternative definition is the steady-state probability that a component or system is out of service. Mathematically, unavailability = (1 – availability).

2.1.4 System Reliability Indexes. The basic system reliability indexes [2], [3], which have proven most useful and meaningful in power distribution system design are:

(1) Load interruption frequency

(2) Expected duration of load interruption events

These indexes can be readily computed using the methods which will be described later. The two basic indexes of interruption frequency and expected interruption duration can be used to compute other indexes which are also useful:

(1) Total expected (average) interruption time per year (or other time period)

(2) System availability or unavailability as measured at the load supply point in question

(3) Expected demanded, but unsupplied, energy per year

It should be noted here that the disruptive effect of power interruptions is often non linearly related to the duration of the interruption. Thus, it is often desirable to compute not only an overall interruption frequency but also frequencies of interruptions categorized by the appropriate durations.

2.1.5 Data Needed for System Reliability Evaluations. The data needed for quantitative evaluations of system reliability depend to some extent on the nature of the system being studied and the detail of the study. In general, however, data on the performance of individual components together with the times required to perform various switching operations are required.

System component data which are generally required are summarized as follows:

(1) Failure rates (forced outage rates) associated with different modes of component failure

(2) Expected (average) time to repair or replace failed component

(3) Scheduled (maintenance) outage rate of component

(4) Expected (average) duration of a scheduled outage event

If possible, component data should be based on the historical performance of components in the same environment as those in the proposed system being studied. The reliability surveys conducted by the Power Systems Reliability

Subcommittee [1], [4] provide a source of component data when such specific data are not available. These data have been summarized in Chapter 3.

The needed switching time data include:

(1) Expected times to open and close a circuit breaker

(2) Expected times to open and close a disconnect or throwover switch

(3) Expected time to replace a fuse link

(4) Expected times to perform such emergency operations as cutting in clear, installing jumpers, etc

Switching times should be estimated for the system being studied based on experience, engineering judgment, and anticipated operating practice.

2.1.6 Method for System Reliability Evaluation. The method for system reliability evaluation which is recommended and presented here has evolved over a number of years [2], [5], [8]. The method, called the *minimal-cut-set method*, is believed to be particularly well suited to the study and analysis of electric power distribution systems as found in industrial plants and commercial buildings. The method is systematic and straightforward and lends itself to either manual or computer computation. An important feature of the method is that system weak points can be readily identified, both numerically and non numerically, thereby focusing design attention on those sections of the system that contribute most to service unreliability. See Chapter 8 for a derivation of the *minimal cut-set-*method.

The procedure for system reliability evaluation is outlined as follows:

(1) Assess the service reliability requirements of the loads and processes which are to be supplied and determine the appropriate service interruption definition or definitions.

(2) Perform a failure modes and effects analysis identifying and listing those component failures and combinations of component failures that result in service interruptions and which constitute minimal cut-sets of the system.

(3) Compute the interruption frequency contribution, the expected interruption duration, and the probability of each of the minimal cut-sets of step (2).

(4) Combine the results of step (3) to produce system reliability indexes.

These steps will be discussed in more detail in the sections that follow.

2.1.7 Service Interruption Definition. The first step in any electric power system reliability study should be a careful assessment of the power supply quality and continuity required by the loads which are to be served. This assessment should be summarized and expressed in a service interruption definition which can be used in the succeeding steps of the reliability evaluation procedure. The interruption definition specifies, in general, the reduced voltage level (voltage dip) together with the minimum duration of such a reduced voltage period which results in substantial degradation or complete loss of function of the load or process being served. Frequently reliability studies are conducted on a *continuity* basis in which case interruption definitions reduce to a minimum duration specification with voltage assumed to be zero during the interruption.

A further discussion of interruption definitions together with examples of such definitions is given in 7.1.2.

2.1.8 Failure Modes and Effects Analysis. Failure modes and effects analysis (FMEA) for power distribution systems amounts to the determination and listing of those component outage events or combinations of component outages which result in an interruption of service at the load point being studied according to the interruption definition which has been adopted. This analysis must be made considering the different types and modes of outages which components may exhibit and the reaction of the system's protection scheme to these events. Component outages may be categorized as:

(1) Forced outages or failures
(2) Scheduled or maintenance outages
(3) Overload outages

Forced outages or failures may be further categorized as:

(1) Permanent forced outages
(2) Transient forced outages

Permanent forced outages require repair or replacement of the failed component before it can be restored to service, while transient forced outages imply no permanent damage to the component, thus permitting its restoration to service by a simple reclosing or refusing operation. Additionally, component failures may be categorized by physical mode or type of failure. This type of failure categorization is important for circuit breakers and other switching devices where failure modes such as the following are possible and have varying impacts on system performance:

(1) Faulted, must be cleared by backup devices
(2) Fails to trip when required
(3) Trips falsely
(4) Fails to re-close when required

The primary result of the failure modes and effects analysis (FMEA) as far as quantitative reliability evaluation is concerned is the list of minimal cut-sets it produces. The use of the minimal cut-sets in the calculation of system reliability indexes is described in Chapter 3 of this book. A minimal cut-set is defined to be a set of components which, if removed from the system, results in loss of continuity to the load point being investigated and which does not contain as a subset any set of components which is itself a cut-set of the system. In the present context the components in a cut-set are just those components whose overlapping outage results in an interruption according to the interruption definition adopted.

An important non quantitative benefit of the failure modes and effects analysis (FMEA) is the thorough and systematic thought process and investigation that it requires. Often weak points in system design will be identified before any quantitative reliability indexes are computed. Thus, the failure modes and affects analysis (FMEA) is a useful reliability design tool even in the absence of the data needed for quantitative evaluation.

The FMEA and the determination of minimal cut-sets is most efficiently conducted by considering first the effects of outages of single components and then the effects of overlapping outages of increasing numbers of components. Those cut-sets containing a single component are termed first-order cut-sets. Similarly cut-sets containing two components are termed second-order cut-sets, etc. In theory the FMEA should continue until all the minimal cut-sets of the system have been found. In practice, however, the FMEA can be terminated earlier since high-order cut-sets have low probability compared to lower order cut-sets. A good rule of thumb is to determine minimal cut-sets up to order $n + 1$ where n is the lowest order minimal cut-set of the system. Since most power distribution systems have at least some first-order minimal cut-sets, the analysis can usually be terminated after the second-order minimal cut-sets have been found.

2.1.9 Computation of Quantitative Reliability Indexes. Computation of reliability indexes may proceed once the minimal cut-sets of the system have been found. The first step is to compute the frequency, expected duration, and expected down time per year of each minimal cut-set. Note that the expected down time per year is just the product of the frequency expressed in terms of events per year and the expected duration. If the expected duration is expressed in years, the expected down time will have the unit of years per year and may be regarded as the relative proportion of time or the probability the system is down due to the minimal cut-set in question. More commonly the expected duration is expressed in hours, and the expected down time has the unit of hours per year.

Approximate expressions for the frequency and expected duration of the most commonly considered interruption events associated with first-, second- and third-order minimal cut-sets are given in Table 1. These are discussed in more detail in Chapter 8.
Note that expressions are given for forced outages (failures) and for forced outages overlapping a maintenance outage. The basic assumptions made in deriving the expressions of Table 1 are as follows:

(1) Component failure and repair events are statistically independent.

(2) Component *up* times are much larger than *down* times. That is, the probability that a component is up is much larger than the probability that it is down.

(3) Components are not taken out of service for maintenance or other deferrable work if other components are on forced outage, but forced outages of components may occur during the scheduled outage of a component.

(4) Components which act in parallel to carry load are fully redundant. That is, any one component of the parallel combination is capable of carrying the entire load of the parallel combination without overload. (Methods for treating overload outages are given in 2.1.12).

Once the frequencies and expected durations have been computed for each minimal cut-set, the system reliability indexes at the load point in question are determined.

f_s = interruption frequency

$$= \sum_{\substack{\text{min} \\ \text{cut-sets}}} f_{cs_i}$$

r_s = expected interruption duration

$$= \sum_{\substack{\text{min} \\ \text{cut-sets}}} f_{cs_i} \, r_{cs_i} / f_s$$

$f_s r_s$ = total interruption time per time period

2.1.10 Example: Sample System Evaluation . The reliability evaluation method which has been described will now be illustrated through a simple example. More detailed examples using typical data are given in Chapter 7.

Consider the simple system of Fig 1. Here, for the sake of simplicity, only the components labeled 1, 2, and 3 will be considered fallible, and only permanent forced outages and scheduled outages

Table 1
Frequency and Expected Duration Expressions for Interruptions Associated with Minimal Cut-Sets

Forced Outages

First-Order Minimal Cut-Set

$$f_{cs} = \lambda_i$$

$$r_{cs} = r_i$$

Second-Order Minimal Cut-Set

$$f_{cs} = \lambda_i \lambda_j (r_i + r_j)$$

$$r_{cs} = r_i r_j / (r_i + r_j)$$

Third-Order Minimal Cut-Set

$$f_{cs} = \lambda_i \lambda_j \lambda_k (r_i r_j + r_i r_k + r_j r_k)$$

$$r_{cs} = r_i r_j r_k / (r_i r_j + r_i r_k + r_j r_k)$$

Forced Outages Overlapping Scheduled Outage

Second-Order Minimal Cut-Set

$$r_{cs} = \left[\lambda_i' \lambda_j' \left(\frac{r_i' r_j'}{r_i' + r_j'} \right) + \lambda_j' \lambda_i r_j' \left(\frac{r_j' r_i}{r_j' + r_i} \right) \right] / f_{cs}$$

Third-Order Minimal Cut-Set

$$f_{cs} = A + B + C$$

$$r_{cs} = \left[A \left(\frac{r_i' r_j r_k}{r_i' r_j + r_i' r_k + r_j r_k} \right) + B \left(\frac{r_j' r_i r_k}{r_j' r_i + r_j' r_k + r_i r_k} \right) + C \left(\frac{r_k' r_i r_j}{r_k' r_i + r_k' r_j + r_i r_j} \right) \right] / f_{cs}$$

where:

$$A = \lambda_i' \lambda_j \lambda_k r_i' \left(\frac{r_i' r_j}{r_i' + r_j} + \frac{r_i' r_k}{r_i' + r_k} \right)$$

$$B = \lambda_j' \lambda_i \lambda_k r_j' \left(\frac{r_j' r_i}{r_j' + r_i} + \frac{r_j' r_k}{r_j' + r_k} \right)$$

$$C = \lambda_k' \lambda_i \lambda_j r_k' \left(\frac{r_k' r_i}{r_k' + r_i} + \frac{r_k' r_j}{r_k' + r_j} \right)$$

Symbols:

f_{cs} = frequency of cut-set event

r_{cs} = expected duration of cut-set event

λ_i = forced outage rate of *i*th component

λ_i' = scheduled outage rate of *i*th component

r_i = expected repair or replacement time of *i*th component

r_i' = expected scheduled outage duration of *i*th component

Note 1: The time units of r and λ in expressions for f_{cs} must be the same.

Note 2: If service can be restored to the affected load point by a switching operation, set r_{cs} = expected switching time. Note that this assumes that switching times are short compared to repair or replacement times.

Fig 1
Sample System

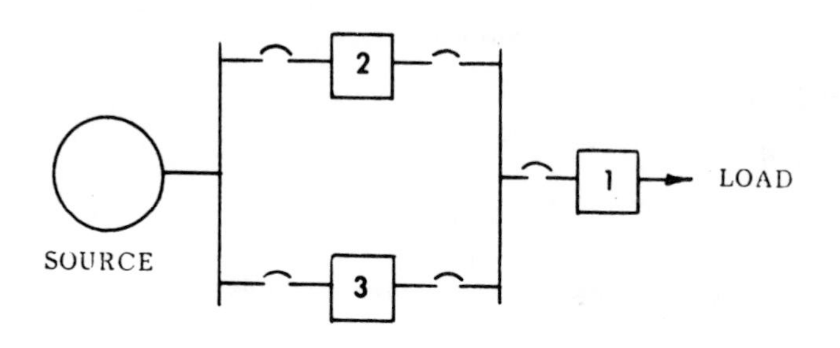

Table 2
Sample System Evaluation

Minimal Cut-Sets	f_{cs}	r_{cs}	$f_{cs}r_{cs}$
	Forced Outages:		
Component 1	λ_1	r_1	$\lambda_1 r_1$
Component 2/Component 3	$\lambda_3\lambda_2(r_3 + r_2)$	$r_3 r_2(r_3 + r_2)$	$\lambda_3\lambda_2 r_3 r_2$
$\left[\lambda_3'\lambda_2 r_3'\left(\frac{r_3' r_2}{r_3'+r_2}\right) + \lambda_2'\lambda_3 r_2'\left(\frac{r_2' r_3}{r_2'+r_3}\right)\right]$	Forced/Scheduled Outages:		
Component 2/Component 3	$\lambda_3'\lambda_2 r_3' + \lambda_2'\lambda_3 r_2'$	$\lambda_3'\lambda_2 r_3' + \lambda_2'\lambda_3 r_2'$	$\lambda_3'\lambda_2 r_3'\left(\frac{r_3' r_2}{r_3'+r_2}\right) + \lambda_2'\lambda_3 r_2'\left(\frac{r_2' r_3}{r_2'+r_3}\right)$
Totals	$\sum f_{cs}$		$\sum f_{cs} r_{cs}$

System indices:

$$f_s = \sum f_{cs}$$

$$r_s' = \sum f_{cs} r_{cs} / f_s$$

will be assumed possible. Additionally, scheduled outages of component 1 which would result in interruptions to the load are assumed to be taken during periods of plant maintenance shutdown and are not treated in the reliability analysis. All circuit breakers are assumed to be normally closed and are coordinated to minimize system disruption due to a component failure.

The load is assumed not to be sensitive to voltage dips limited to the time required to clear a fault, but any loss of continuity from source to load constitutes an interruption to the load. Using this definition of an interruption, the minimal cut-sets of the system are shown in Table 2. Also shown in Table 2 are the expressions needed for the calculation of the system reliability indexes.

2.1.11 Post Fault Switching. Power systems are frequently designed with protective and switching schemes that allow restoration of service through some switching action following one or more component failures. The key point in such cases is that the duration of the interruption event is the time to perform the switching operation rather than the time to repair or replace failed components. Thus when computing the frequency and duration of a system interruption event (minimal cut-set), the frequency is computed as outlined in Tables 1 and 2, but the duration becomes simply the switching time (assuming that the switching time is short compared to the repair or replacement time).

The procedure is most easily described through use of a simple example. Consider the system of Fig 2 in which power is supplied from a source to a load over two parallel circuits with the sectionalizing scheme shown. Assume for the purposes of the example that only the circuits are subject to failure and that each circuit has a failure rate of λ_i and an average repair time of r_i. Assume further that the sectionalizing switches at the load point can be opened in a time t_s to isolate a faulty circuit and that all circuit breakers and switches are normally closed.

If the definition of system failure is loss of continuity from source to load, the general reliability evaluation procedure yields the results given in Table 3.

Table 3
Sample System Evaluation
Considering Post-Fault Switching

Minimal-Cut	Frequency	Duration	Probability
Failure of circuit 1	λ_1	t_s	$\lambda_1 t_s$
Failure of circuit 2	λ_2	t_s	$\lambda_2 t_s$
Failure of both circuits	$\lambda_1\lambda_2(r_1 + r_2)$	$\dfrac{r_1 r_2}{(r_1 + r_2)}$	$\lambda_1\lambda_2 r_1 r_2$

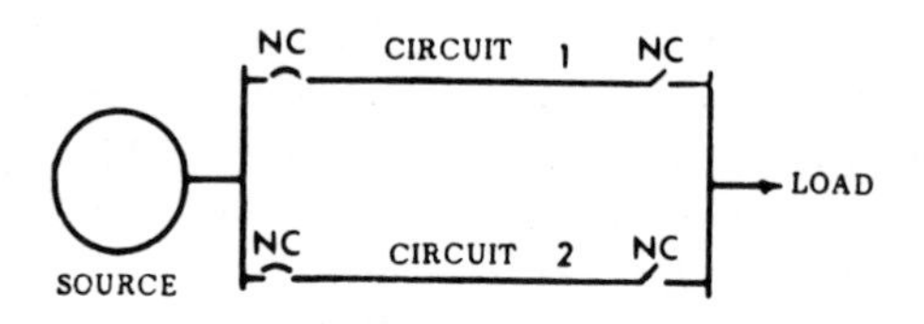

Fig 2
Example of Post–Fault Switching

Note that *failure of both circuits* is not, strictly speaking, a minimal cut-set since failure of either circuit alone constitutes a system failure and load interruption. The justification for including *failure of both circuits* is the desire to give frequency, duration, and probability information on both short-duration interruptions which are terminated by switching actions and long-duration interruptions which are terminated by repair or replacement efforts. In the event that load sensitivity is such that differences in power supply interruption durations are not important, *failure of both circuits* could have been omitted and only the frequency of minimal cut-sets calculated. In general system failure, frequency and probability are obtained by summing the frequency and probability columns, respectively. If it is desired to separate short- and long-duration interruption performance, these summations are carried out with due regard to the average durations of the various cut-sets. Average interruption duration is given by system failure probability divided by system failure frequency.

2.1.12 Incomplete Redundancy. A common method of improving the reliability performance of a system is through component redundancy, for example, more than one transformer in a substation. Typically, each component of the redundant set has sufficient capacity, perhaps based on an emergency rating, to carry the peak load that the system may be asked to deliver. Such full redundancy is effective in improving system reliability performance but is usually quite expensive. If the load of the system is variable, the opportunity exists to cut costs by reducing the capacity of redundant components to levels less than that required to carry system peak load. Such component capacity reductions admit the possibility that one component of the redundant set might be called upon to carry system peak load and would thereby suffer an overload outage. An overload outage might result in an actual interruption of load or perhaps only some loss of life in the overloaded component, depending on the protection scheme in service.

A method exists [9], [10] for computing the frequency, average duration,

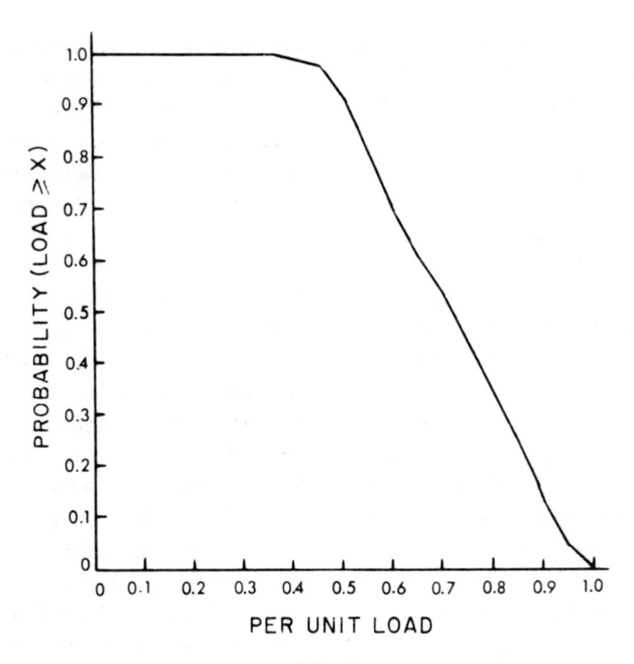

Fig 3
Typical Load–Duration Characteristic

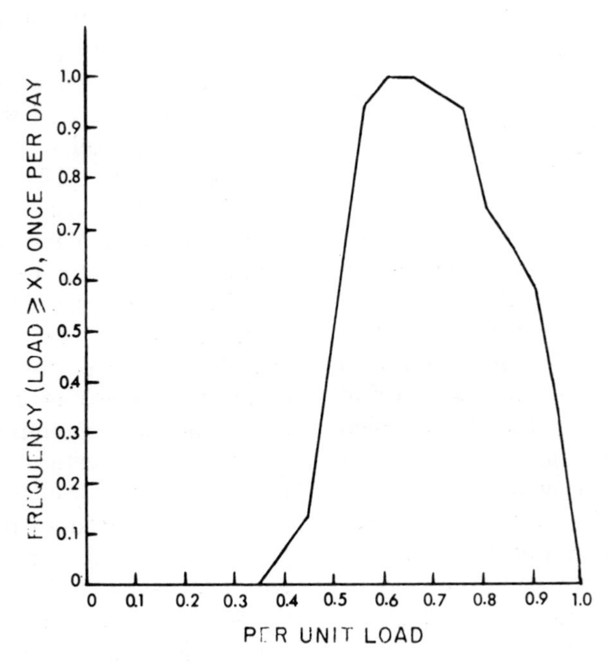

Fig 4
Typical Load–Frequency Characteristic

and probability of overload outage events as a function of component capacities and load characteristics. This method, which is compatible with the general reliability evaluation procedure outlined earlier, can be used to evaluate the cost–reliability tradeoffs of incomplete redundancy. The method is briefly presented hereafter.

Consider a system possessing incomplete redundancy, and consider the forced outage of some set i of the components of this system. Let the frequency and probability of this forced outage event be f_i and P_i. Then the frequency, probability, and average duration of overloading events that are precipitated by loss of the components in set i are given approximately by:

$$f_{OL_i} = f_i \cdot P \text{ (load} \geqslant \text{capacity of remaining components)} + P_i \cdot f \text{ (load} \geqslant \text{capacity of remaining components)}$$

$$P_{OL_i} = P_i \cdot P \text{ (load} \geqslant \text{capacity of remaining components)}$$

$$D_{OL_i} = P_{OL_i} / f_{OL_i}$$

In the above expressions P (load $\geqslant X$) is called the load–duration characteristic and is simply the probability of proportion of time that the load is greater than or equal to X. A typical load–duration characteristic for utility load is shown in Fig 3. Similarly, f (load $\geqslant X$) is called the load–frequency characteristic and is the rate with which events (load $\geqslant X$) occur. A typical load–frequency characteristic is shown in Fig 4. The reader is referred to [10] for additional discussion of the load–duration and load–frequency characteristics.

2.2 Costs of Interruptions; Economic Evaluation of Reliability

2.2.1 Cost of Interruptions Versus Capital Cost. The type and extent of new or rehabilitated electric systems for industrial plants or commercial buildings must carefully balance the costs of anticipated interruptions to electrical service against the capital costs of the systems involved. Each instance requires a separate analysis taking into account special production and occupancy needs. Because of the many variables involved, one of the most difficult items to obtain is the cost of the electrical interruptions.

2.2.1.1 What is an Interruption? Economic evaluation of reliability begins with the establishment of an interruption definition. Such a definition specifies the magnitude of the voltage dip and the minimum duration of such a reduced-voltage period which result in a loss of production or other function for the plant, process, or building in question. Frequently, interruption definitions are given only in terms of a minimum duration and assume that the voltage is zero during that period.

IEEE surveys, [12], [13], [14], have revealed a wide variation in the minimum or *critical* service loss duration. Table 4 summarizes results for industrial plants and Table 5 gives results for commercial buildings. It is clear from these tables that careful attention must be paid to choosing the proper interruption definition in any specific reliability evaluation.

Another important consideration in the economic evaluation of reliability is the time required to restart a plant or process following a power interruption. An IEEE survey [12], (shown in Table 6) indicates that industrial plant restart time following a complete plant shut-

Table 4
Critical Service Loss Duration for Industrial Plants*

(Maximum length of time an interruption of electrical service will not stop plant production.)

25th Percentile	Median	75th Percentile	Average Plant Outage Time for Equipment Failure Between 1- and 10-Cycle Duration
10.0 cycles	10.0 seconds	15 minutes	1.39 hours

*55 plants in the United States and Canada reporting; all industry.

Table 5
Critical Service Loss Duration for Commercial Buildings*

(Maximum length of time before an interruption to electrical service is considered critical.)

Service Loss Duration Time							
1 cycle %	2 cycles %	8 cycles %	1 second %	5 minutes %	30 minutes %	1 hour %	12 hours %
3	6	9	15	36	64	74	100

*Percentage of buildings with critical service loss for duration less than or equal to time indicated (54 buildings reporting).

down due to a power interruption averages 17.4 hours. The median plant restart time was found to be 4.0 hours. Clearly, specific data on plant or process restart time should be used if possible in any particular evaluation.

Many industrial plants reported that 1 to 10 cycles were considered critical interruption time, as compared to 1.39 h, required for startup, (plant outage time being considered equal to plant startup time). This indicates that the critical factor must be carefully explored prior to assigning a cost to the interruption. That 15% of the commercial buildings reported the critical service loss duration time to be 1 second or less is probably attributable to the fact that computer installations were involved.

Further data from [12] graphically illustrates the time required to start an industrial plant after an interruption.

Step 1 of the cost analysis thus becomes the selection of the *critical dura-*

Table 6
Plant Restart Time*

(After service is restored following a failure that has caused a complete plant shutdown.)

Average (hours)	Median (hours)
17.4	4.0

*43 plants in the United States and Canada reporting; all industry.

tion time of the outage and the *plant startup* time, including equipment repair or replacement time required because of the interruption.

2.2.12 Cost of an Electrical Service Interruption. With the establishment of expected downtime per interruption, costs are assigned to all individual items involved, including but not limited to:

(1) Value of loss production time less expenses saved (expected restart time is used along with the repair or replacement time)

(2) Damaged plant equipment

(3) Spoiled or off-specification product

(4) Extra maintenance costs

(5) Cost for repair of failed component

If possible, the cost for each interruption of service should be expressed in a short interruption plus an amount of dollar per hour for the total outage time in order to utilize the reliability data and analysis presented.

2.2.1.3 Economic Evaluation of Reliability . There are many methods of varying degrees of complexity for accomplishing economic evaluations. For quick *order-of-magnitude* or *Is it worth further investigation?* type of evaluations, cost data from [12], [14] can be used. Caution must be exercixed however, since these data are very general in nature, and wide variations are possible in individual cases. Some of the more commonly accepted methods for economic analyses are:

(1) Revenue requirements (RR)

(2) Return on investment (ROI)

(3) Life cycle costing (LCC)

It is not the intent to stipulate here the method to be used nor the depth to which each analysis is to be made. These are considered to be the prerogative of the engineer and will depend heavily on management choice and the time available for the analysis. The revenue requirements (RR) method is given in this chapter as an example.

2.2.2 *"Order of Magnitude" Cost of Interruptions.* IEEE surveys, References [2], [4], presented general data on the cost of interruptions to industrial plants and commercial buildings. The reader is again cautioned that such general data should be used only for "order of magnitude" evaluations where data specific to the system being studied is not available. A review of the reliability data can probably best be used in adjudicating the type of utility company service which should be provided.

The costs based on the kW interrupted and the kWh not delivered to industrial plants are presented in Tables 7 and 8.

Interruption costs based on kWh not delivered and reflecting the relationship to duration of interruptions for commercial buildings are presented in the Tables 9 and 10.

2.2.3 *Economic Analysis of Reliability in Electrical Systems.* There are several acceptable methods for accomplishing an economic analysis of the reliability in electric systems. The examples of reliability analysis included in this chapter and Chapter 7 utilize the Revenue Requirements (RR) method. The application of this method as it applied to the analyses of the reliability in Industrial Plant Electrical Systems was presented in Reference part 6 of [2]. Applicable excerpts from this reference are included herein.

2.2.3.1 *The Revenue Requirements (RR) Method.* Although there are many ways in use to compare alternatives, some of these have defects and weaknesses, especially when comparing design alternatives in contrast to overall projects. The RR method is "mathematically rigorous and quantitatively correct to the extent permitted by accuracy with which items of cost can be forecast." [2], [13]

The essence of the RR method is that for each alternative plan being considered, the Minimum Revenue Requirements (MRR) is determined. This means that we find out how much product we must sell to achieve minimum acceptable earnings on the investment involved plus all expenses associated with that investment. These MRRs for alternative plans may be compared directly. The plan having the lowest MRR is the economic choice.

MRR is made up of and equal to (1) variable operating expenses, (2) minimum acceptable earnings, (3) depreciation, (4) income taxes, and (5) fixed operating expenses. These MRRs may be separated into two main parts, one proportional and the other not propor-

Table 7
Average Cost of Power Interruptions for Industrial Plants*

All Plants	\$1.89/kW + \$2.68/kWh
Plants > 1000 kW Max Demand	\$1.05/kW + \$0.94/kWh
Plants < 1000 kW Max Demand	\$4.59/kW + \$8.11/kWh

*41 plants in the United States and Canada reporting—published in 1973.

Table 8
Median Cost of Power Interruptions for Industrial Plants*

All plants	$0.69/kW + $0.83/kWh
Plants > 1000 kW Max Demand	$0.32/kW + $0.36/kWh
Plants < 1000 kW Max Demand	$3.68/kW + $4.42/kWh

*41 plants in the United States and Canada reporting—published in 1973.

Table 9
Average Cost of Power Interruptions for Commercial Buildings

All commercial buildings*	$7.21/kWh not delivered
Office buildings only	$8.86/kWh not delivered

*54 buildings in the United States reporting—published in 1975.

Table 10
Cost of Power Interruptions
as a Function of Duration
for Office Buildings (with Computers)*

Power Interruptions	Sample Size	Maximum	Cost/Peak kWh Not Delivered Minimum	Average
15-min duration	14	$22.22	$1.88	$8.89
1-Hour duration	16	$24.93	$1.88	$8.30
Duration > 1-hour	10	$67.66	$0.16	$9.81

*Published in 1975.

tional to investment in the alternative. This may be expressed in an equation:

$$G = X + CF \qquad \text{(Eq 1)}$$

where

G = minimum revenue requirements (MRR) to achieve minimum acceptable earnings

X = nonfixed or variable operating expenses

C = capital investment

F = fixed investment charge factor

The last term in Eq 1, the product of C and F includes the items (2), (3), (4), and (5) listed in the preceding paragraph. Equation 1 is now discussed.

X—Variable Expenses. The effect of the failure of a component is to cause an increase in variable expenses. How serious this increase is depends to a great extent on the location of the component in the system and on the type of power distribution system employed. The quality of a component as installed can have a significant effect on the number of failures experienced. A poor quality component installed with poor workmanship and with poor application engineering may greatly increase the number of failures that occur as compared with a high quality component installed with excellent workmanship and sound application engineering.

When a failure does occur, variable expenses are increased in two ways. In the first way, the increase is the result of the failure itself. In the second way, the increase is proportional to the duration of the failure.

Considering the first way, the increased expense due to the failure includes the following:

(1) Damaged plant equipment

(2) Spoiled or off-specification product

(3) Extra maintenance costs

(4) Costs for repair of the failed component

Considering the second way, plant downtime resulting from failures is made up of the time required to restart the plant, if necessary, plus the time to (1) effect repairs, if it is a radial system, or (2) effect a transfer from the source on which the failure occurred to an energized source.

During plant downtime, production is lost. This lost production is not available for sale, so revenues are lost. However, during plant downtime, some expenses may be saved, such as expenses for material, labor, power, and fuel costs. Therefore, the value of the lost production is the revenues lost, because production stopped less the expenses saved. Some of the variable expenses may vary depending on the duration of plant downtime. For example, if plant downtime is only one hour, perhaps no labor costs are saved. But, if plant downtime exceeds eight hours, labor costs may be saved.

If we assume that the value/hour of variable expenses does not vary with the duration of plant downtime, then the value of lost production can be expressed on a per hour basis, and the total value of lost production is the product of plant downtime in hours and the value of lost production per hour.

It should be noted that both the value of lost production and expenses incurred are proportional to the failure rate. The total effect on variable expenses, if the value of lost production is a constant on a per hourly basis, may be expressed in an equation.

$$X = \lambda [x_i + (g_p - x_p)(r+s)] \qquad \text{(Eq 2)}$$

where

X = variable expenses ($ per year)

λ = failures per year or failure rate

x_i = extra expenses incurred per failure ($ per failure)

g_p = revenues lost per hour of plant downtime ($ per hour)

x_p = variable expenses saved per hour of plant downtime ($ per hour)

r = repair or replacement time after a failure (or transfer time if not radial system), hours

s = plant startup time after a failure, hours

Assume that

λ = 0.1 failure per year

x_i = $20,000 per failure, extra expenses incurred

g_p = $8,000 per hour, revenues lost

x_p = $6,000 per hour, expenses saved

r = 10 hours per failure

s = 20 hours per failure

Then, variable expenses affected would be:

$$X = (0.1)[\$20,000 + (\$8,000 - \$6,000)(10 + 20)]$$
$$= \$8,000 \text{ per year}$$

The term g_p represents revenues lost and it is not really an expense. However, it is a negative revenue, and as such has the same effect on the economics as a positive expense item. It is convenient to treat it as though it were an expense.

A failure rate of 0.1 failure per year is equivalent to a mean time between failures of 10 years. Since we are dealing with probability, this is what we can expect, but in a specific case, we might have two failures in one 10 year period and no failures in another 10 year period. But considering many similar cases, we expect to have an average of 0.1 failure per year, with each failure costing an average of $80,000. This gives an equal average amount per year in the above example of $8,000.

The point is that even though the actual failures cost $80,000 each and occur once every 10 years, a given failure is just as likely to occur in any of the 10 years. The equivalent equal annual amount of $8,000 per year is the average value of one failure in 10 years.

c—Investment. Each different alternative is an industrial plant power distribution system involves different investments. The system requiring the least investment will usually be some form of radial system. By varying the type of construction and the quality of the components in the system, the investment in radial systems can vary widely.

The best method is to find one total investment in each alternative plan. Another common method is to find the incremental investment in all alternatives over a base or least expensive plan. The main reason that the total investment method is preferable, is that in comparing alternatives, the investment is multiplied by an F factor which will be explained later. This factor is usually the same for alternative plans of the sort being considered here, but this is not necessarily the case.

Using the incremental investment may thus introduce a slight error into the economic comparisons.

F—Investment Charge Factor. This discussion of investment charge factor is taken from [2].

The factor F includes the following items which are fixed in relation to the investment.

(1) Minimum acceptable rate of return on investment, allowing for risk

(2) Income taxes

(3) Depreciation

(4) Fixed expenses

An equation to calculate the F factor is:

$$F = \left[\frac{\left(S_c a_L / f_r\right) - t\overline{d_t}}{1-t} \right] + e \qquad \text{(Eq 3)}$$

This may also take the following form:

$$F = \overline{r} + \overline{d} + \overline{t} + e \qquad \text{(Eq 4)}$$

where

a_n = $R + d_n$, amortization factor or leveling factor

d_n = $R/(S_n-1)$, sinking fund factor

S_n = $(1 + R)^n$, growth factor or future value factor

n = period of years, such as c or L

c = years prior to startup that an investment is made

L = life of investment years

R = minimum acceptable earnings per \$ of C, investment

f_r = probability of success or risk adjustment factor

t = income taxes per \$ of C, investment

$\overline{d}_t$ = income tax depreciation, levelized per \$ of C, investment = 1/L , d_t = 1L

e = fixed expenses per \$ of C, investment

$\overline{r}$ = levelized return on investment per \$ of C, investment

$\overline{d}$ = levelized depreciation on investment per \$ of C, investment

$\overline{t}$ = levelized income taxes on investment per \$ of C, investment

Assume

L = 20 years, life of investment

c = 1 year

R = 0.15, minimum acceptable rate of return

f_r = 1, risk adjustment factor

t = 0.5, income tax rate

$\overline{d}_t = \dfrac{1}{L} = 0.05$

e = 0.0825

Then

$S_c = (1+R)^c = (1+0.15)^1 = 1.15$

$S_L = (1+R)^L = (1+0.15)^{20} = 16.37$

$d_L = R/(S_L-1) = 0.15/(16.37-1) = 0.0098$

$a_L = R+d_L = 0.15 + 0.0098 = 0.1598$

Substituting into Eq 3 to calculate the F factor, we get:

$$F = \left[\frac{\frac{(1.15)(0.1598)}{1.0} - (0.5)(0.05)}{(1-0.5)} \right] + 0.0825 = 0.4 \qquad \text{(Eq 5)}$$

All the assumed values are believed to be typical for the average electric distribution system, except the value of e = 0.0825. This latter value was arbitrarily assumed to make R round out to 0.4. The term e covers such items as insurance, property taxes, and fixed maintenance costs. A typical value is probably less than 0.0825.

It is believed that a typical value for minimum acceptable return on investment in many industrial plants is 15 percent, that is, $R = 0.15$. The company average rate of return, based on either past history or anticipated results, is a measure of what R should be. In plants of higher risk than the average, the risk adjustment factor, f_r, should probably be less than 1. However, company management determines what the value of R should be.

The value of F can be calculated from Eq 3 using a log log slide rule. In [14], tabular values are given for the factors S_n and a_n for various rates of return and plant lifes.

2.2.3.2 *Steps for Economic Comparisons.*

1. Prepare one-line diagrams of alternative plans and assign failure rates, repair times, and investment in each component, and determine the total investment C, in each plan.
2. Determine X, the increased variable expense for each plan as the sum of the value of lost production and the extra variable expenses incurred.
3. Determine F, the fixed investment charge factor, F from Eq 3.
4. Calculate G = X + CF, the minimum revenue requirements G of each plan. Eq 1.
5. Select as the economic choice the plan having the lowest value of G.

2.2.3.3 *Conclusions.* A technique has been presented for the economic evaluation of power system reliability. The method of determining the failure rates and repair times of different alternatives is not covered here. Additional information relative to the RR Method is included in [13].

2.2.4 *Examples.* Examples of electric systems with varying degrees of reliability (availability), together with fixed and variable costs are given in Chapter 7.

2.3 Cost of Scheduled Electrical Preventive Maintenance

It is always appropriate to consider in the economic evaluation of reliability the costs for scheduled electrical preventive maintenance. Sometimes these costs are large enough to make it desirable to analyze them separately when comparing alternative designs of industrial power systems. The *Revenue Requirements Method* described in 2.2.3 includes a term called the Investment Charge Factor, F which is given by Eq 3 in 2.2.3 and includes e the fixed yearly expenses e (as a percentage of the capital investment) are attributed to scheduled electrical preventive maintenance, insurance, property taxes, etc. Since the yearly average costs for scheduled electrical preventive maintenance may not be the same percentage of investment for every component within the industrial power system, a separate more detailed look is often taken at these costs for each component.

Scheduled electrical preventive maintenance has two major cost elements; labor effort and spare parts consumed. These costs are often expressed on an average yearly basis so as to be usable with the "Revenue Requirements Method" when an economic evaluation is made. These data are needed for each different type of component used in the industrial power system and can be compiled for each component as follows:

(1) Labor Costs in manhours per component per year
(2) Cost of spare parts consumed in dollars per component per year
(3) Labor rate in dollars per manhour

If, for example, a component is only maintained once every three years, then its maintenance costs should be divided by three in order to determine the average yearly maintenance cost. The labor rate used probably should only include the overhead costs associated with the storage of spare parts, direct supervision of the maintenance, and costs for necessary test equipment. The labor costs in dollars per component per year can be calculated by multiplying items (1) and (3) together; the result can then be added to item (2) to get the total average yearly costs that are attributable to scheduled electrical preventive maintenance.

Data thus collected can become obsolete at a later date due to inflation which can result in changing the labor rate used and also the average yearly cost of spare parts consumed. But the data for labor in manhours per component per year does not become obsolete due to inflation. Some engineers have chosen to use their labor rate to convert their average yearly cost data for spare parts consumed into average yearly "equivalent manhours" data. This is then added to the labor manhours data to get total equivalent manhours per component per year that includes both the labor cost and the cost of spare parts consumed. The use of equivalent manhours for cost data instead of dollars has two advantages:

(1) The equivalent manhours data do not become obsolete due to inflation

(2) The equivalent manhours data can be considered an international currency The data are not affected by changing exchange rates between the currencies or different countries. This enables the cost data to be compared with studies from other countries.

No component data are included in this book on the cost of scheduled electrical preventive maintenance. It would be desirable to have such data for all of the electrical equipment categories listed in Table 11 of Chapter 3. It would then be possible to consider the cost of scheduled electrical preventive maintenance in design decisions of the industrial power system by adding this into the Minimum "Revenue Requirements Method."

2.4 Effect of Scheduled Electrical Preventive Maintenance on Failure Rate

One of the important total operating cost decisions made by the management of an industrial plant is how much money to spend for scheduled electrical preventive maintenance. The amount of maintenance performed on a component can affect its failure rate. Very little quantitative data have been collected and published on this subject. Yet this is an important factor when attempting to study the total owning costs of a complete power system. If maintenance

Table 11
Summary of All-Industry Equipment Failure Rate and Equipment Outage Duration Data for 60 Equipment Categories Containing Eight or More Failures

(See Tables 4 through 19 in Appendix A for additional details)

Equipment	Equipment Subclass	Failure Rate (failures per unit-year)	Actual Hours of Downtime Per Failure	
			Industry Average	Median Plant Average
Transformers	Liquid filled — all	0.0041	529.0	219.0
	601–15 000 V — all sizes	0.0030	*174.0	49.0
	300–750 kVA	0.0037	61.0	10.7
	751–2499 kVA	0.0025	217.0	64.0
	2500 kVA and up	0.0032	216.0	60.0
	Above 15 000 V	0.0130	*1076.0	1260.0
	Dry type: 0–15 000 V	0.0036	153.0	28.0
	Rectifier: Above 600 V	0.0298	380.0	80.0
Circuit breakers	Fixed type (including molded case) — all	0.0052	5.8	4.0
	0–600 V — all sizes	0.0044	4.7	4.0
	0–600 A	0.0035	2.2	1.0
	Above 600 A	0.0096	9.6	8.0
	Above 600 V	0.0176	10.6	3.8
	Metalclad drawout — all	0.0030	129.0	7.6
	0–600 V — all sizes	0.0027	*147.0	4.0
	0–600 A	0.0023	3.2	1.0
	Above 600 A	0.0030	232.0	5.0
	Above 600 V	0.0036	*109.0	168.0
Motor starters	Contact type: 0–600 V	0.0139	65.1	24.5
	Contact type: 601–15 000 V	0.0153	284.0	16.0
Motors	Induction: 0–600 V	0.0109	*114.0	18.3
	Induction: 601–15 000 V	0.0404	*76.0	91.5
	Synchronous: 0–600 V	0.0007	35.3	35.3
	Synchronous: 601–15 000 V	0.0318	*175.0	153.0
	Direct current — all	0.0556	37.5	16.2
Generators	Steam turbine driven	0.032	165.0	66.5
	Gas turbine driven	0.638	23.1	92.0
Disconnect switches	Enclosed	0.0061	3.6	2.8
Switchgear bus — indoor and outdoor (Unit = number of connected circuit breakers or instrument transformer compartments)	Insulated: 601–15 000 V**	0.00170	261.0	26.8
	Bare: 0–600 V**	0.00034	550.0	24.0
	Bare: Above 600 V**	0.00063	17.3	13.0
Bus duct — indoor and outdoor (Unit = 1 circuit ft)	All voltages	0.000125	128.0	9.5
Open wire (Unit = 1000 circuit ft)	0–15 000 V	0.0189	42.5	4.0
	Above 15 000 V	0.0075	17.5	12.0
Cable — all types of insulation (Unit = 1000 circuit ft)	Above ground and aerial			
	0–600 V	0.00141	457.0	10.5
	601–15 000 V — all	0.01410	*40.4	6.9
	In trays above ground	0.00923	8.9	8.0
	In conduit above ground	0.04918	140.0	47.5
	Aerial cable	0.01437	31.6	5.3

Table 11 (Continued)

Equipment	Equipment Subclass	Failure Rate (failures per unit-year)	Actual Hours of Downtime per Failure: Industry Average	Actual Hours of Downtime per Failure: Median Plant Average
Cable — all types of insulation (Unit = 1000 circuit ft)	Below ground and direct burial			
	0–600 V	0.00388	15.0	24.0
	601–15 000 V — all	0.00617	*95.5	35.0
	In duct or conduit below ground	0.00613	96.8	35.0
	Above 15 000 V	0.00336	16.0	16.0
Cable (Unit = 1000 circuit ft)	601–15 000 V			
	Thermoplastic	0.00387	44.5	10.0
	Thermosetting	0.00889	168.0	26.0
	Paper insulated lead covered	0.00912	48.9	26.8
	Other	0.01832	16.1	28.5
Cable joints — all types of insulation	601–15 000 V			
	In duct or conduit below ground	0.000864	36.1	31.2
Cable joints	601–15 000 V			
	Thermoplastic	0.000754	15.8	8.0
	Paper insulated lead covered	0.001037	31.4	28.0
Cable terminations — all types of insulation	Above ground and aerial			
	0–600 V	0.000127	3.8	4.0
	601–15 000 V — all	0.000879	198.0	11.1
	Aerial cable	0.001848	48.5	11.3
	In trays above ground	0.000333	8.0	9.0
	In duct or conduit below ground			
	601–15 000 V	0.000303	25.0	23.4
Cable terminations	601–15 000 V			
	Thermoplastic	0.004192	10.6	11.5
	Thermosetting	0.000307	451.0	11.3
	Paper insulated lead covered	0.000781	68.8	29.2
Miscellaneous	Inverters	1.254	107.0	185.0
	Rectifiers	0.038	39.0	52.2

*See Tables 48 through 56 in Appendix B for results of a special study on *Effect of Failure Repair Method and Failure Repair Urgency on the Average Hours Downtime per Failure.*

**See Appendix E for data from a later survey.

NOTE: The following tables in Appendix A contain reliability data for equipment categories that had a small sample size (4–7 failures):

Table 6	Circuit breakers used as motor starters
Table 8	Generators driven by motor, diesel, or gas engine
Table 9	Disconnect switches — open
Table 15	Cable joints, 601–15 000 V, above ground and aerial
Table 16	Cable joints, 601–15 000 V, thermosetting
Table 19	Fuses
Table 19	Protective relays

effort is reduced the maintenance costs go down. This may increase the failure rate of the components in the power system and raise the costs associated with failures. There is an optimum amount of maintenance for minimum total owning cost of a complete power system.

The subject of Electrical Preventive Maintenance is discussed in Chapter 5. Some data are shown in Tables 21 and 22 on the effect of the frequency and quality of scheduled electrical preventive maintenance. These data have been used to calculate the effect of maintenance quality on the failure rate of transformers, circuit breakers, and motors shown in Table 23. Unfortunately the data do not relate the amount or cost of component maintenance to the failure rate.

The effect of the cost of component scheduled electrical preventive maintenance on the failure rate has not been included in this book. More industry studies and published data are needed on this subject.

2.5 References[2]

[1] IEEE COMMITTEE REPORT, Report on Reliability Survey of Industrial Plants, *IEEE Transactions on Industry Applications*, March/April 1974, pp 213–235.

[2] DICKINSON, W. H. et al, Fundamentals of Reliability Techniques as Applied to Industrial Power Systems, *Conf Record 1971 IEEE I & CPS Technical Conference*, pp 10–31.

[3] PATTON, A. D. and AYOUB, A. K. Reliability Evaluation, *Systems Engineering for Power: Status and Prospects*, U.S. Energy Research and Development Administration, publication CONF-750867, 1975, pp 275–289.

[4] IEEE COMMITTEE REPORT, Reliability of Electric Utility Supplies to Industrial Plants, *Conf Record 1975 I & CPS Technical Conference*, pp 131–133.

[5] GAVER, D. P., MONTMEAT, F. E., and PATTON, A. D. Power System Reliability, I—Measures of Reliability and Methods of Calculation, *IEEE Trans on PA & S*, July 1964, pp 727–737.

[6] RINGLEE, R. J. and GOODE, S. D. On Procedures for Reliability Evaluation of Transmission Systems, *Ibid*, April 1970, pp 527–537.

[7] ENDRENYI, J., MAENHAUT, P. C., PAYNE, L. C. Reliability Evaluation of Transmission Systems with Switching After Faults—Approximations and a Computer Program, *Ibid*, pp 1963–1875, Nov/Dec 1973.

[8] BILLINTON, R. and GROVER, M. S. A Sequential Method for Reliability Analysis of Distribution and Transmission Systems, *Proceedings 1975 Annual Reliability and Maintainability Symposium*, Jan 1975, pp 460–469.

[9] CHRISTIAANSE, W. R., Reliability Calculations including the Effects of Overloads and Maintenance, *IEEE Trans on PA & S*, July/Aug 1971, pp 1664–1676.

[10] AYOUB, A. K., and PATTON, A. D. A Frequency and Duration Method for Generating System Reliability Evaluation, IEEE Trans Power App Syst, Nov/Dec 1976, pp 1929–1933.

[11] IEEE COMMITTEE REPORT, Report on Reliability Survey of Industrial Plants, *IEEE Transactions on Industry Applications*, July/Aug 1975.

[2] References [1], [4], [11], and [12] are reprinted in Appendixes A, B, C, and D.

[12] PATTON, A. D. et al, Cost of Electrical Interruptions in Commercial Buildings, *IEEE I & CPS Conference Record* of May 5-8, 1975.

[13] JEYNES, P. H. and VAN NEMWEGEN, L. The Criterion of Economic Choice, Trans AIEE (Power Apparatus and Systems), vol 77, August 1958, pp 606–635.

[14] DICKINSON, W. H. Economic Evaluation of Industrial Power Systems Reliability, *Trans AIEE (Applications and Industry)* Vol 76, Nov 1957, pp 264–272.

3. Summary of Equipment Reliability Data

3.1 Introduction

A knowledge of the reliability of electrical equipment is an important consideration in the design of power distribution systems for industrial plants and commercial buildings. Ideally these reliability data should come from field use of the same type of equipment under similar environmental conditions and similar stress levels. In addition there should be a sufficient number of field failures in order to represent an adequate sample size. It is believed that eight field failures are the minimum number necessary in order to have a reasonable change of determining a failure rate or an "average downtime per failure" to within a factor of two. The types of reliability data needed on each component of electrical equipment are:

(1) Failure rate—failures per year

(2) Average downtime to repair or replace a piece of equipment after a failure—hours (or minutes) per failure

(3) Information on pertinent factors than can have an effect on (1) or (2) above.

These reliability data on each component of electrical equipment can then be used to represent historical experience for use in cost-reliability and cost-availability tradeoff studies in the design of new power distribution systems.

From 1973 to 1975 the Power Systems Reliability Subcommittee of the IEEE Industrial Power Systems Department conducted and published [1], [1] [2] surveys of electrical equipment reliability in industrial plants. These reliability surveys of electrical equipment and electric utility power supplies were extensive, and summaries of pertinent reliability data are given in this chapter.

(1) Failure rate and outage duration time for electrical equipment and electric utility power supplies

(2) Failure characteristic or failure modes of electrical equipment; that is, the effect of the failure on the system

(3) Causes and types of failures of electrical equipment

[1]Numbers in brackets correspond to those in the References at the end of this Section.

(4) Failure repair method and failure repair urgency

(5) Method of service restoration after a failure

(6) Loss of motor load versus time of power outage.

In addition reference is made to summaries of pertinent reliability data and information that are contained in other chapters. This includes:

(7) Maximum length of time an interruption of electrical service will not stop plant production.

(8) Plant restart time after service is restored following a failure that caused a complete plant shutdown.

(9) Cost of power interruptions to industrial plants and commerical buildings.

(10) Example showing that the two power sources in a double-circuit utility supply are not completely independent.

(11) Equipment failure rate multipliers versus maintenance quality.

(12) Percentage of failures caused by inadequate maintenance versus month since maintained

All of the reliability data summarized in the above twelve items are taken from the IEEE surveys of industrial plants [1] [2] and commercial buildings [3]. The detailed reports are given in Appendixes A, B, C, and D. A later survey [4] of the reliability of switchgear bus is included in Appendix E.

3.2 Reliability of Electrical Equipment

The pertinent failure rate and average downtime per failure information for the in-plant electrical equipment are given in Table 11. In compiling these data, a failure was defined as any trouble with a power system component that causes any of the following effects:

(1) Partial or complete plant shutdown, or below-standard plant operation

(2) Unacceptable performance of user's equipment

(3) Operation of the electrical protective relaying or emergency operation of the plant electric system

(4) Deenergization of any electric circuit or equipment

A failure of an in-plant component causes a forced outage of the component,and the component thereby is unable to perform its intended function until it is repaired or replaced.

All of the 60 electrical equipment categories listed in Table 11 have at least 8 or more failures. This is considered an adequate sample size in order to have a reasonable chance of determining a failure rate within a factor of two. Failure rate and average downtime per failure data for an additional seven categories of equipment are contained in [1] (Appendix A.). These additional categories of equipment have between 4 to 7 failures and might be considered by some as too small a sample size, they include:

(1) Circuit breakers used as motor starters

(2) Generators driven by motor, diesel, or gas engine

(3) Disconnect switches—open

(4) Cable joints, 601–15,000 V, above ground and aerial

(5) Cable joints, 601–15,000 V, thermosetting

(6) Fuses

(7) Protective relays

The "failure repair method" and the "failure repair urgency" have a significant effect on the "average downtime per failure" for some equipment categories. Special studies have been made of this subject for nine equipment categories that are marked with an asterisk (*) in Table 11. These studies are reported in Tables 48 through 56 of [1] (Appendix B).

3.3 Reliability of Electric Utility Power Supplies to Industrial Plants

The "failure rate" and the "average downtime per failure" of electric utility supplies to industrial plants are given in Table 12. Additional details are given in [2] (Appendix D). A total of 87 plants participated in the IEEE survey covering the period from January 1, 1968 through October 1974.

The survey results shown in Table 12 have distinguished between power failures that were terminated by a switching operation versus repair or replacement of equipment. The latter have a much longer outage duration time. Some of the conclusions that can be drawn from these IEEE data are:

(1) The failure rate for single-circuit supplies is about six times that of multiple circuit supplies which operate with all circuit breakers closed; and the average duration time of each outage is about 2.5 times as long

(2) Failure rates for multiple-circuit supplies which operate with either a manual or an automatic throwover scheme are comparable to those for single-circuit supplies, but throwover schemes have a smaller average failure duration than single-circuit supplies

(3) Failure rates are highest for utility supply circuits operated at distribution voltages and lowest for circuits operated at transmission voltages

It is important to note that the data in Table 12 show that the two power sources of a double-circuit utility supply are not completely independent. This is analyzed in an example in 7.1.15, where for the one case analyzed the actual failure rate of a double-circuit utility supply is more than 200 times larger than the calculated value for two completely independent utility power sources. See Table 36 of Chapter 7 for a summary table comparing the actual and calculated failure rates of a double-circuit utility power supply.

It is believed that utility supply failure rates vary widely in various locations. One of the significant factors in this difference is believed to be different exposures to lightning storms. Thus, average values for the utility supply failure rate may not be valid to use at any one location. Local values should be obtained, if possible, from the utility involved, and these values should be used in reliability and availability studies.

An earlier IEEE reliability survey of electric utility power supplies to industrial plants was published in 1973 and is reported in Table 3 of [1] (Appendix A). The earlier survey had a smaller data base and is not believed to be as accurate nor as up to date as the one summarized in Table 12. The earlier survey of electric utility power supplies had lower failure rates.

3.4 Failure Characteristic of Electrical Equipment

The failure characteristic or failure modes of electrical equipment are important. They identify the effect of the failure on the system and are needed in system reliability studies. Tables 13, 14, and 15 give the failure characteristic of transformers, circuit breakers, and 11 other categories of electrical equipment.

There are some "failure modes" of circuit breakers that require backup protective equipment to operate; for example, "failed to trip" or "failed to inter-

Table 12
IEEE Survey of Reliability of Electric Utility Supplies to Industrial Plants [2]
(See Tables II, III, IV, V in Appendix D for additional details)

Single Circuit Utility Supplies

	Failures per Unit-Year**			Average Duration (minutes per failure**)		
Voltage level	λ_S	λ_R	λ	r_S	r_R	r
V⩽15 kV	0.905	2.715	3.621	3.5	165	125
15 kV<V⩽35 kV	—	1.657	1.657	—	57	57
V>35 kV	0.527	0.843	1.370	1.5	59	37
All	0.556	1.400	1.956	2.3	110	79

Multiple Circuit Utility Supplies
All Voltage Levels

	Failure per Unit-Year**			Average Duration (minutes per failure**)		
Switching scheme	λ_S	λ_R	λ	r_S	r_R	r
All breakers closed	0.255	0.057	0.312	8.5	130	31
Manual throwover	0.732	*0.118	0.850	8.1	*84	19
Automatic throwover	1.025	0.171	1.196	0.6	96	14
All	0.453	0.085	0.538	5.2	110	22

Multiple Circuit Utility Supplies
All Switching Schemes

	Failures per Unit-Year**			Average Duration (minutes per failure**)		
Voltage level	λ_S	λ_R	λ	r_S	r_R	r
V⩽15 kV	0.640	0.148	0.788	4.7	149	32
15 kV<V⩽35 kV	0.500	*0.064	0.564	4.0	*115	17
V>35 kV	0.357	0.067	0.424	6.1	184	34

Multiple Circuit Utility Supplies
All Circuit Breakers Closed

	Failures per Unit-Year**			Average Duration (minutes per failure**)		
Voltage level	λ_S	λ_R	λ	r_S	r_R	r
V⩽15 kV	0.175	*0.088	0.263	0.7	*335	112
15 kV<V⩽35 kV	0.342	*0.019	0.361	7.0	*120	13
V>35 kV	0.250	0.061	0.311	11.0	203	49

*Small sample size — less than 8 failures.

**Failure rates λ and average durations r subscripted S and R are, respectively, rates and durations of failures terminated by switching and by repair or replacement. Unsubscripted rates and durations are overall values.

Table 13
Failure Characteristic of Transformers

%	Failure Characteristic
92	Automatic removal by protective equipment
1	Partial failure reducing capacity
7	Manual removal

rupt." Both of these failure modes would require that a circuit breaker further up the line be opened, and this opening would result in a larger part of the power distribution system being disconnected. Thus the data on failure modes of circuit breakers shown in Table 14 are important in system reliability studies.

The data in Table 15 show that a significant percentage of the failures of

Table 14
Failure Modes of Circuit Breakers
(Percentage of Total Failures in Each Failure Mode)

All Circuit Breakers	Metalclad Drawout			*Fixed Type		
	All	601–15 000 V	0–600 V All Sizes	0–600 V All Sizes	All	
%	%	%	%	%	%	Failure Characteristic
5	5	2	7	8	6	Failed to close when it should
9	12	21	0	0	2	Failed while opening
42	58	49	71	5	4	Opened when it should not
7	6	4	9	5	4	Damaged while successfully opening
2	1	0	0	0	4	Damaged while closing
32	16	24	10	77	73	Failed while in service (not while opening or closing)
1	0	0	0	0	2	Failed during testing or maintenance
1	2	0	3	0	0	Damage discovered during testing or maintenance
1	0	0	0	5	5	Other
100%	100%	100%	100%	100%	100%	Total percentage
165	117	53	59	39	48	Number of failures in total percentage
8	7	0	7	1	1	Number not reported
173	124	53	66	40	49	Total failures

*Includes molded case.

motor starters, motors, and disconnect switches are discovered during testing or maintenance.

3.5 Causes and Types of Failures of Electrical Equipment

Data are given in Tables 16 and 17 on:

(1) Failures, damaged part

(2) Failure type

(3) Suspected failure responsibility

(4) Failure initiating cause

(5) Failure contributing cause

The data in Table 17 indicate that indequate maintenance is suspected as responsible for a significant percentage of the failures for several categories of electrical equipment.

3.6 Equipment Failure Rate Multipliers Versus Maintenance Quality

The relationship between maintenance practice and equipment failure is discussed in 5.3.

Equipment failure rate multipliers versus maintenance quality are given in Table 23 of Chapter 5 for transformers, circuit breakers, and motors. These multipliers were determined in a special study part 6 of [1]; (Appendix B). The failure rate of motors is very sensitive to the quality of maintenance.

The percentage of failures due to inadequate maintenance versus the months since maintained is given in Table 22 of Chapter 5 for circuit breakers, motors, open wire, transformers, and all electrical equipment classes combined. A high percentage of equipment failures are blamed on inadequate maintenance if there has been no maintenance for more than two years prior to the failure.

Table 15
Failure Characteristics of Other Electrical Equipment

Motor Starters	Motors	Generators	Disconnect Switches	Switchgear Bus — Insulated	Switchgear Bus — Bare	Bus Duct	Open Wire	Cable	Cable Joints	Cable Terminations	Failure Characteristic
%	%	%	%	%	%	%	%	%	%	%	
37	69	74	72	100	87	90	68	92	96	80	Failed in service
5	1	0	3	0	13	5	2	2	4	2	Failed during testing or maintenance
36	30	0	18	0	0	0	1	2	0	12	Damage discovered during testing or maintenance
20	0	16	6	0	0	5	6	3	0	6	Partial failure
2	0	10	1	0	0	0	23	1	0	0	Other

Table 16
Failure, Damaged Part, and Failure Type

Trans-formers	Circuit Breakers	Motor Starters	Motors	Gen-erators	Dis-connect Switches	Switch-gear Bus — Insulated	Switch-gear Bus — Bare	Bus Duct	Open Wire	Cable	Cable Joints	Cable Termi-nations	
%	%	%	%	%	%	%	%	%	%	%	%	%	Failure, Damaged Part
68	0	5	50	7	0	0	0	15	0	5	0	0	1. Insulation — winding
13	2	0	0	0	1	5	8	10	1	0	0	12	2. Insulation — bushing
3	19	10	3	0	14	90	71	65	6	84	91	75	3. Insulation — other
0	1	0	29	2	0	0	0	0	0	3	0	0	4. Mechanical — bearings
0	11	16	3	7	9	0	0	0	0	0	0	0	5. Mechanical — other moving parts
1	6	2	1	4	30	0	0	0	4	1	0	4	6. Mechanical — other
3	6	13	3	10	8	5	0	0	3	1	0	0	7. Other electrical — auxiliary device
1	28	2	0	1	1	0	0	0	3	1	0	0	8. Other electrical — protective device
7	1	0	0	0	0	0	0	0	0	0	0	0	9. Tap changer — no load type
1	0	0	0	0	0	0	0	0	0	0	0	0	10. Tap changer — load type
3	26	52	11	69	38	0	21	10	84	6	9	10	99. Other
													Failure Type
58	33	14	28	19	15	65	79	70	34	73	70	55	1. Flashover or arcing involving ground
13	10	20	4	3	4	35	21	30	23	1	9	4	2. All other flash-over or arcing
12	19	55	32	29	47	0	0	0	25	7	20	37	3. Other electrical defect
10	11	11	31	32	14	0	0	0	6	5	0	4	4. Mechanical defect
7	27	0	6	16	21	0	0	0	12	14	0	0	99. Other

3.7 Failure Repair Method and Failure Repair Urgency

The "failure repair method" and the "failure repair urgency" have a significant effect on the "average downtime per failure." Table 18 shows the percentages of these two parameters for thirteen classes of electrical equipment. A special study on this subject is reported in Tables 48 through 56 of [1] (Appendix B) for nine categories of electrical equipment (marked with an * in Table 11 of this chapter).

The IEEE data on "method of electrical service restoration to plant" are shown in Table 19. A percentage breakdown of the total shows:

Replacement of failed component with spare	22%
Repair of failed component	22%
Other	22%
Utility service restored	12%
Secondary selection—manual	11%
Primary selection—manual	7%
Primary selection—automatic	2%
Secondary selection—automatic	2%
Network protector operation-automatic	0+%

The most common methods of service restoration are replacement of failed component with a spare or repair of failed component. Only 22 percent of the time is primary selection or secondary selection used. This would indicate that most power distribution systems in the IEEE survey were radial.

3.8 Loss of Motor Load Versus Time of Power Outage

A special study has been reported in Table 47 of [1] (Appendix B) on loss of motor load versus time of power outage. For power outages longer than 10 cycle duration, most plants lose motor load. However, for power outages between 1 and 10 cycle duration, only about half as many plants lose the motor load.

Test results on the effect of fast bus transfers are reported in [5]. This includes 4 kV induction and synchronous motors with the following types of loads: (1) forced draft fan, (2) circulating water pump, (3) boiler feed booster pump, (4) condensate pump, (5) gas recirculation fan. A list of prior papers on the effect of fast bus transfer on motors is also contained in [5].

3.9 Critical Service Loss Duration Time

What is the maximum length of time that an interruption of electrical service will not stop plant production? The median value for all plants is 10.0 seconds. See Table 4 in Chapter 2 for a summary of the IEEE survey of industrial plants.

What is the maximum length of time before an interruption to electrical service is considered critical in commerical buildings? The median value of all commerical buildings is between 5 and 30 minutes. See Table 5 in Chapter 2 for a summary of the IEEE survey of commercial buildings.

3.10 Plant Restart Time

What is the plant restart time after service is restored following a failure that has caused a complete plant shutdown? The median value for all plants is 4.0 hours. See Table 6 in Chapter 2 for a summary of the IEEE survey of industrial plants.

Table 17
Suspected Failure Responsibility, Failure Initiating Cause, and Failure Contributing Cause

Transformers	Circuit Breakers	Motor Starters	Motors	Generators	Disconnect Switches	Switchgear Bus — Insulated	Switchgear Bus — Bare	Bus Duct	Open Wire	Cable	Cable Joints	Cable Terminations	
%	%	%	%	%	%	%	%	%	%	%	%	%	Suspected Failure Responsibility
39	23	18	15	19	29	5	9	26	0	16	0	0	1. Manufacturer—defective component
0	0	0	0	0	0	0	0	0	0	0	0	0	2. Transportation to site — defective handling
2	4	51	9	0	6	45	4	16	2	8	0	18	3. Application engineering — improper application
3	3	0	1	3	4	10	17	5	9	14	50	39	4. Inadequate installation and testing prior to start-up
11	23	8	17	19	13	35	22	16	30	10	18	22	5. Inadequate maintenance
9	6	3	4	3	40	0	0	0	2	3	0	0	6. Inadequate operating procedures
2	5	0	0	0	1	0	22	5	5	4	5	0	7. Outside agency — personnel
4	1	0	1	6	0	0	17	0	21	6	2	8	8. Outside agency — other
30	36	19	53	48	8	5	9	32	31	38	25	14	9. Other
													Failure Initiating Cause
23	13	1	6	10	4	5	5	0	26	26	11	12	1. Transient overvoltage disturbances (lightning, switching surges, arcing ground fault in ungrounded system)

Table 17 (Continued)

Transformers	Circuit Breakers	Motor Starters	Motors	Generators	Disconnect Switches	Switchgear Bus — Insulated	Switchgear Bus — Bare	Bus Duct	Open Wire	Cable	Cable Joints	Cable Terminations	Failure Initiating Cause
%	%	%	%	%	%	%	%	%	%	%	%	%	
0	0	0	0	0	0	0	0	0	0	0	0	0	2. Overvoltage
11	3	1	26	3	4	0	5	30	21	1	0	2	3. Overheating
18	18	8	30	3	5	50	18	20	8	29	40	51	4. Other insulation breakdown
17	13	8	4	29	17	10	23	45	7	24	31	24	21. Mechanical breaking, cracking, loosening, abrading or deforming of static or structural parts
0	5	6	20	3	2	0	0	0	0	0	0	0	22. Mechanical burnout, friction, or seizing of moving parts
1	1	0	3	3	20	0	0	0	10	7	0	4	23. Mechanically caused damage from foreign source (digging, vehicular, accident, etc)
1	2	5	0	0	0	0	23	5	14	2	0	2	41. Shorting by tools or metal objects
2	1	1	0	0	0	0	9	0	3	0	0	2	42. Shorting by birds, snakes, rodents, etc
0	1	0	0	3	0	0	0	0	0	0	0	0	51. Loss of control power
1	11	64	5	0	0	0	0	0	0	0	0	0	52. Malfunction of protective relay control device, or auxiliary device
0	0	0	0	0	3	0	0	0	0	0	0	0	61. Low voltage
0	0	0	0	0	0	0	0	0	0	0	0	0	62. Low frequency
25	33	7	5	45	45	35	18	0	11	10	18	4	99. Other

Table 17 (Continued)

Trans-formers	Circuit Breakers	Motor Starters	Motors	Gen-erators	Dis-connect Switches	Switch-gear Bus — Insulated	Switch-gear Bus — Bare	Bus Duct	Open Wire	Cable	Cable Joints	Cable Termi-nations	Failure Contributing Cause
%	%	%	%	%	%	%	%	%	%	%	%	%	
13	4	0	5	10	8	0	0	6	0	2	0	0	1. Persistent overloading
0	1	0	1	6	3	5	0	0	0	0	2	0	2. Above-normal temperatures
0	0	0	0	0	1	0	0	0	0	0	0	0	3. Below-normal temperature
0	2	0	7	0	0	0	10	0	28	14	13	10	4. Exposure to aggressive chemicals or solvents
6	3	0	10	6	4	15	20	17	1	8	22	12	5. Exposure to abnormal moisture or water
0	0	0	0	3	0	0	5	0	3	2	0	0	6. Exposure to non-electrical fire or burning
0	0	0	2	0	0	0	0	0	0	1	0	0	8. Obstruction of ventilation by foreign objects or material
24	17	40	34	32	5	20	10	50	3	30	29	24	9. Normal deterioration from age
6	1	0	2	3	0	20	5	11	30	15	2	16	10. Severe wind, rain, snow, sleet, or other weather conditions
0	2	0	0	6	0	0	0	0	1	0	0	0	11. Protective relay improperly set
0	1	2	15	0	0	0	0	0	0	0	0	0	12. Loss or deficiency of lubricant
0	0	0	1	0	0	0	0	0	0	0	0	0	13. Loss or deficiency of oil or cooling medium
3	10	3	0	0	0	0	15	6	2	3	0	8	14. Misoperation or testing error
3	3	1	5	0	26	40	20	0	2	1	0	0	15. Exposure to dust or other contaminants
44	55	53	18	32	53	0	15	11	31	24	31	29	99. Other

Table 18
Failure Repair Method and Failure Repair Urgency

Transformers	Circuit Breakers	Motor Starters	Motors	Generators	Disconnect Switches	Switchgear Bus — Insulated	Switchgear Bus — Bare	Bus Duct	Open Wire	Cable	Cable Joints	Cable Terminations	
%	%	%	%	%	%	%	%	%	%	%	%	%	**Failure Repair Method**
47	51	33	78	84	30	95	71	65	70	47	87	60	1. Repair of failed component in place or sent out for repair
53	49	67	22	16	70	5	29	35	9	53	13	34	2. Repair by replacement of failed component with spare
0	0	0	0	0	0	0	0	0	21	0	0	6	99. Other
													Failure Repair Urgency
51	73	66	23	48	20	70	58	80	55	66	56	53	1. Requiring round-the-clock all out efforts
45	22	34	74	52	80	25	33	15	26	28	22	31	2. Requiring repair work only during regular workday, perhaps with some overtime
4	5	0	2	0	0	5	8	5	0	6	22	16	3. Requiring repair work on a nonpriority basis
0	0	0	0	0	0	0	0	0	19	0	0	0	99. Other

Table 19
Method of Service Restoration

Total	Electric Utility Power Supplies	Transformers	Circuit Breakers	Motor Starters	Motors	Generators	Disconnect Switches	Switchgear Bus — Insulated	Switchgear Bus — Bare	Bus Duct	Open Wire	Cable Joints	Cable joints	Cable Terminations	
%	%	%	%	%	%	%	%	%	%	%	%	%	%	%	
7	1	3	6	0	5	20	0	58	25	20	13	14	28	19	1. Primary selection — manual
2	8	0	1	0	0	0	0	0	5	0	4	5	8	0	2. Primary selection — automatic
11	1	25	6	0	14	33	0	17	10	10	2	20	32	23	3. Secondary selection — manual
2	1	3	8	0	0	0	0	0	0	0	1	0	8	4	4. Secondary selection — automatic
0+	0	0	0	0	0	0	0	0	5	0	0	0	0	0	5. Network protector operation — automatic
22	5	25	11	12	30	20	3	17	20	35	31	42	24	27	6. Repair of failed component
22	2	39	38	10	29	14	77	0	10	35	6	2	0	12	7. Replacement of failed component with spare
12	81	0	1	0	0	13	0	0	0	0	1	1	0	0	8. Utility service restored
22	1	5	29	78	22	0	20	8	25	0	42	16	0	15	9. Other
100	100	100	100	100	100	100	100	100	100	100	100	100	100	100	Total percentage
1204	171	75	160	68	318	15	69	12	20	20	103	122	25	26	Total number reported

3.11 Cost of Power Interruptions to Industrial Plants and Commercial Buildings

The cost of power interruptions to industrial plants is summarized in Tables 7 and 8 of Chapter 2.

The cost of power interruptions to commercial buildings is summarized in Tables 9 and 10 of Chapter 2. The effect of computer outages on the costs is shown in these tables.

3.12 Reliability Improvement of Electrical Equipment in Industrial Plants Between 1962 and 1973

The failure rates of electrical equipment in industrial plants have improved considerably during the eleven-year interval between the 1962 AIEE reliability survey [6] and the 1973 IEEE reliability survey [1]. Table 20 shows how much the failure rates have improved for several equipment categories. These data are calculated from a 1974 report [7]. In 1962 circuit breakers had failure rates that were 2.5 to 6.0 times higher than in 1973. The largest improvements in equipment failure rates have occurred on generators, cables, motors, and circuit breakers. The authors discussed some of the reasons for the failure rate improvements during the eleven-year interval. It would appear that manufacturers, application engineering, installation engineering, and maintenance personnel have all contributed to the overall reliability improvement.

The authors also make a comparison the "actual downtime per failure" for all equipment categories shown in the table. In general, the "actual downtime per failure" was larger in 1973 than in 1962.

Table 20
Failure Rate Improvement Factor of Electrical Equipment in Industrial Plants During 11 Year Interval Between 1962 AIEE Survey and 1973 IEEE Survey

Equipment Category	Failure Rate Ratio AIEE (1962) / IEEE (1973)
Generators	
Steam turbine driven	16.0
Gas turbine driven**	13.0
Cable	
Nonleaded in underground conduit	9.7
Nonleaded, aerial	5.8
Lead covered in underground conduit	3.4
Nonleaded in above-ground conduit	1.6
Cable joints and terminations	
Non-leaded	5.3
Leaded	2.0
Motors, 50 hp and larger***	
Synchronous	8.4
Induction	7.1
Circuit breakers	
Metalclad drawout, 0–600 V	6.0
Metalclad drawout, Above 600 V	2.9
Fixed 2.4–15 kV	2.5
Disconnect switches	
Open, above 600 V	3.4
Enclosed, above 600 V	1.6
Open wire	3.4
Transformers	
Below 15 kV, 0–500 kVA****	2.0
Below 15 kV, above 500 kVA	2.0
Above 15 kV	1.6
Motor starters, contactor type	
0–600 V	1.3
Above 600 V	1.3

*Unit = 1000 circuit ft.

**Middle East pipeline data only for 1962.

***250 hp and larger for 1962.

****300–750 kVA for 1973.

3.13 Other Sources of Reliability Data

The reliability data from industrial plants which are summarized in this chapter are based upon References [1] [2] and [3] which were published during 1973-1975. Reference [6] is an earlier reliability survey of industrial plants which had been published in 1962.

Many sources of reliability data on similar types of electrical equipment exist in the electric utility industry. The Edison Electric Institute (EEI) has collected and published reliability data on power transformers, power circuit breakers, metal-clad switchgear, motors, excitation systems, generators, and prime mover generation equipment [8]—[14]. Most EEI reliability activities do not collect outage duration time data; only Reference [4] on prime mover generation equipment contains such data.

Failure rate data and outage duration time data of power transformers, power circuit breakers and buses are given in References [5] and [6]. These data have come from electric utility power systems.

Very little published data are available on the failure modes of power circuit breakers and on the probability of a circuit breaker not operating when called upon to do so. An extensive worldwide reliability survey of the major failure modes of power circuit breakers above 63 kV on utility power systems has been made by CIGRE 13-06 Working Group [7]. Failure rate data and failure per operation data have been determined for each of the major failure modes. Outage duration time data have also been collected. In addition data have been collected on the costs of scheduled preventive maintenance; this includes the manhours per circuit breaker per year and the cost of spare parts consumed per circuit breaker per year.

ANSI/IEEE Std 500 1977 [17], is a reliability data manual for use in the design of nuclear power generating stations. The equipment failure rates cover such equipment as annunciator modules, batteries and chargers, blowers, circuit breakers, switches, relays, motors and generators, heaters, transformers, valve operators and actuators, instruments, controls, sensors, cables, raceways, cable joints, and terminations. No information is included on equipment outage duration times. Future revisions to ANSI/IEEE Std 500-1977 will include mechanical component data.

The Edison Electric Institute also sponsors the Nuclear Plant Reliability Data System (NPRDS) which collects failure data on electrical components in the safety systems of nuclear power plants. Outage duration time data are collected on each failure. It is expected that NRPDS will become a source of reliability data for future revisions to ANSI/IEEE Std 500-1977.

3.14 References[2]

[1] IEEE COMMITTEE REPORT, Report on Reliability Survey of Industrial Plants, Parts 1–6 *IEEE Transactions on Industry Application*, March/April, July/August, September/October 1974, pp 213–252, 456–476, 681.

[2] IEEE COMMITTEE REPORT, Reliability of Electric Utility Supplies to Industrial Plants, IEEE—ICPS Technical Conference, Toronto, Canada, May 5–8, 1975, pp 131–133 in 75-CHO947-1-1A.

[2] References [2] and [3] are reprinted in Appendixes A, B, C, and D.

[3] IEEE COMMITTEE REPORT, Cost of Electrical Interruptions in Commercial Buildings, IEEE—ICPS Technical Conference, Toronto, Canada, May 5–8, 1975, pp 124–129 in 75–CHO947–1–1A.

[4] See Appendix E—IEEE COMMITTEE REPORT, Report of Switchgear Bus Reliability Survey of Industrial Plants, *IEEE Transactions on Industry Applications,* Mar/Apr 1979, pp 141–147).

[5] AVERILL, E. L. Fast Transfer Test of Power Station Auxiliaries, *IEEE Transactions on Power Apparatus and Systems,* Vol PAS–96, pp 1004–1009, May/June 1977.

[6] DICKINSON, W. H. Report of Reliability of Electrical Equipment in Industrial Plants, *AIEE Transactions,* Part II, pp 132–151,July 1962.

[7] McWILLIAMS, D. W., PATTON, A. D., HEISING, C. R. Reliability of Electrical Equipment in Industrial Plants—Comparison of Results from 1959 Survey and 1971 Survey, IEEE—ICPS Technical Conference, Denver, Jun 2–6, 1974, pp 105–112 in 74CHO855–71A.

[8] Report of Power Transformer Troubles, 1975, *EEI Publication No 76–80,* Dec 1976.

[9] Report on Power Circuit Breaker Troubles, 1975, *EEI Publication No 76–81,* Dec 1976.

[10] Report on Metalclad Switchgear Troubles, 1975, *EEI Publication No 76–82,* Dec 1976.

[11] Report on Motor Troubles, 1975, *EEI Publication No 76–79,* Dec 1976.

[12] Report on Excitation System Troubles, 1975, *EEI Publication No 76–78,* Dec 1976.

[13] Report on Generator Troubles, 1975, *EEI Publication No 76–77,* Dec 1976.

[14] Report on Equipment Availability for Ten Year Period 1965-1974, *EEI Publication No 75–50* (Prime Mover Generation Equipment). These data are now collected and published by National Electric Reliability Council.

[15] PATTON, A. D. Determination and Analysis of Data for Reliability Studies, *IEEE Transactions of Power Apparatus and Systems,* Vol PAS-87 Jan 1968.

[16] MALLARD, S. A. and THOMAS, V. C. *A Method for Calculating Transmission System Reliability,* Vol PAS–87, Mar 1968.

[17] Will be published in *Electra.*

[18] ANSI/IEEE Std 500–1977, IEEE Guide to the Collection and Presentation of Electrical, Electronic, and Sensing Component Reliability Data for Nuclear Power Generating Stations.

4. Evaluating and Improving Reliability of an Existing Plant

4.1 Introduction

The 1974 survey of electrical equipment reliability in Industrial Plants [1][1] and subsequent investigations showed the utility supply as being the largest single component affecting the reliability of an industrial plant. (See Table 3 in Appendix A and Table 12 in Chapter 3.) The industrial user may or may not be in a position to improve the utility supply reliability and, as a result, must also focus his attention on critical areas within his own plant. A logical approach to the analysis of options available in the industrial plant (in terms of both utility supply and plant distribution) will lead to the most reliability improvement for the least cost. In many instances, reliability improvements can be obtained without any cost by making the proper inquiries.

Most industrial users simply "hook up" to the utility system and do not fully recognize that their requirements can have an impact on how the utility supplies them. A utility is somewhat bound by the system available at the plant site and the investment that can be made per revenue dollar. However, most utilities are willing to discuss the various supply systems that are available to their customers. Many times, an option is available (sometimes with financial sharing between the user and the utility) that will meet the exact reliability needs of an industrial plant.

A thorough and properly integrated investigation of the entire electric system (plant and supply) will pinpoint the components or sub systems having unacceptable reliability. Some important general inquiries are listed below. Many of these questions apply to both the utility and the plant distribution systems.

(1) How is the system supposed to operate?

(2) What is the physical condition of the electrical system?

(3) What will happen if faults occur at different points?

[1]Numbers in brackets correspond to those in the References at the end of this Section.

(4) What is the probability of a failure and its duration?

(5) What is the critical duration of a power interruption that will cause significant financial loss? (That is, will a one-minute interruption cost production dollars or merely be an inconvenience?)

(6) Is there any fire or health hazard that will be precipitated by an electrical fault or a power loss?

(7) Is any equipment vulnerable to voltage dips or surges?

The answers to these and similar questions if properly asked can and will result in savings to the industrial user if they produce *action*.

A question at this point should be—How do I get started? However, another question could be—Why bother? The answer to the former question is covered in this chapter, and the answer to the latter question is based on the following analogy. When preparing for a long trip, a motorist will make sure that his car is in good working condition before he leaves. He will check the brakes, engine, transmission, tires, exhaust system, etc., to see that they are in good condition and make the required repairs. For the motorist knows that "on the road" breakdowns and failures are expensive, time consuming, and can be hazardous. In an industrial plant, an unplanned electrical failure will consume valuable production time as well as dollars and may cause injury to personnel. Circuit breakers, relays, meters, transformers, wireways, etc, need periodic checks and preventive maintenance (see Chapter 5) to improve the likelihood of trouble-free performance. Some plants have been shut down completely by events such as a ballast failure. These "shutdowns" are commonly caused by improper settings in protective devices, circuit breaker contacts that were welded shut, or relays that were not set (or did not react) properly. This chapter shows the plant engineer how to minimize downtime by analyzing his system.

4.2 Utility Supply Availability

Loss of incoming power will cause an interruption to critical areas unless alternate power sources are available. Therefore, the reliability of the incoming power is of paramount importance to the plant engineer. It can be stated that different plants and even circuits within a plant vary in their response to loss of power. In some cases, production will not be significantly affected by a 10-minute power interruption. In other cases, a 10-millisecond interruption will cause significant loss. The plant engineer should assess his plant's vulnerability and convey his requirements to the local utility (as well as his own management). See Section 2.2 of Chapter 2 for information on economic loss versus unavailability of incoming power.

The local utility should be able to supply a listing of the number, type and duration of power interruptions over the preceding 3 to 5 year period. The utility should also be able to predict the future average performance based on the historical data and planned construction projects. In addition, the utility may be able to supply the "feeder" performance of other circuits near the facility under investigation. A second alternative would be to obtain a diagram of the utility feed and evaluate its availability using Chapter 2 methods. As a last resort, the average numbers in this book will provide a good base (Table 12 of Chapter 3).

The utility's history of interruptions can be compared with recorded plant dollar loss in verifying the process vulnerability. By assigning a dollar loss to each interruption it will be possible to determine a relationship between the duration of a power loss and a monetary loss for a particular industrial plant. When the actual outage cost is higher or lower than would be predicted, the cause of the deviation should be determined (that is, a 15 minute power loss at a shift change will be less costly than one during peak production). With a refined cost formula in hand, the cost of available options versus projected losses can be evaluated.

Occasionally, a plant will experience problems at times other than during a recorded outage. These problems may be caused by voltage dips (or—more rarely—voltage spikes) which are difficult to trace. With problems such as these, it is necessary to begin recording the exact date and time of these occurrences and ask the utility to search for faults or other system disturbances at (or near) the specific times that they have been recorded. It would be wise to convey the fault times to the utility reasonably soon after the fault (that is, call them the following day). It must be emphasized that unless these problems are significant in terms of dollars lost, safety or frequency (that is, every other day), it is not reasonable to pursue the cause of voltage dips since they are a natural phenomenon in the expansive system operated by a utility. Frequent dips can be caused by large motor starts, welder inrush or intermittent faults in the plant's distribution system (or even by a neighbor's system).

It is also reasonable to cover "what if" questions with the utility and to weigh their answers in any supply decision. A list of questions include:

(1) How long will the plant be without power if:

(a) The main transformer fails

(b) The feed to the main transformer fails

(c) The pole supporting the plant feed is struck by a vehicle and downed

(d) The utility main line fuse or protector interrupts

(e) The utility main feed breaker opens for a fault

(f) The utility substation transformer fails

(g) The utility substation feeds are interrupted

(2) What kind of response time can be expected from the utility for loss of power:

(a) During a lightning storm

(b) During a low trouble period (That is, "normal" conditions.)

(c) During a snow or ice storm

(d) During a heat storm (That is, long periods of high temperatures.)

(3) What should be done when the plant experiences an interruption

(a) Who should be called? A name and number should be made available to *all* responsible personnel. Alternates and their numbers should be included also.

(b) What information should be given to those called?

(c) How should plant people be trained to respond?

(d) Can plant personnel restore power by switching utility lines and who should be contacted to obtain permission to switch?

(4) Are there any better performing feeders near the plant and what is the cost of extending them to the plant?

(That is, is a spare feeder available and what is the cost to make it available?)

(a) Is this additional feed from the same station or from another station?

(b) What is the probability (frequency and duration) of both the main and the spare feeds being interrupted simultaneously?

(c) What is the reliability improvement obtained from the additional (or alternate) feed?

(5) Will the utility protective equipment coordinate with the plant's service circuit breaker? If not, what can be done to coordinate these series protective devices?

(6) What is the available short circuit current, and are there plans to change the system so as to effect the short circuit current?

All of the above questions may not apply to all plants, but should be matched with specific plant requirements.

There is an important fact to consider when a multiple ended feed is being considered. While service is maintained for a loss of one of the feeds, a voltage depression will be seen until the fault is cleared by proper relay action. Therefore, the plant will see a voltage dip for *any* faults on *all* incoming feeds. If the plant is affected with equal severity by either a voltage dip or a short-duration (several seconds) interruption, a multiple ended supply (with secondary tie) may actually worsen plant reliability. This is just one example of the need to carefully evaluate the current supply situation in conjunction with the net improvement of various proposals.

4.3 Where to Begin—The Plant One-Line Diagram

The "blue print" for electrical analysis is the "one line diagram." The existence of a one-line diagram is essential for any plant electrical engineer, manager or operator. It is the "roadmap" to any part of the electric system. In fact, a one-line diagram should be prepared even if the ensuing analysis is not done.

The one-line diagram should begin at the incoming power supply. Standard IEEE symbols should be used in representing electrical components (see ANSI/IEEE Std 315-1975 Graphic Symbols for Electrical and Electronics Diagrams). It is usually impractical to show all circuits in a plant on a single schematic; so the initial one-line diagram should show only major components, circuits, and panels. More detailed analysis may be required in critical areas (described later) and additional one-line diagrams should be prepared for these areas as required.

Since an analysis is being made from the one-line diagram, the type, size and rating of each device as well as its unavailability should be shown on the diagram. The diagram should include at least the following information:

(1) Incoming lines (voltage, size—capacity and rating)

(2) Generators (in plant)

(3) Incoming main fuses, potheads, cutouts, switches, main and tie breakers

(4) Power transformers (rating, winding connection and grounding means)

(5) Feeder breakers and fused switches

(6) Relays (function, use and type)

(7) Potential transformers (size, type and ratio)

(8) Current transformers (size, type and ratio)

(9) Control transformers

All main cable and wire runs with their associated isolating switches and potheads. (Size and length of run).

All substations including integral relays and main panels. Exact nature of load in each feeder and on each substation.

The one-line diagram may show planned, as well as actual, feeder circuit breaker and substation loads (actual measurements should be taken). In most industrial plants, load is added (or deleted) in small increments, and the net effect is not always seen until some part of the system becomes overloaded (or underloaded). Many times, circuits are added without appropriate modification of the standard settings on the associated upstream circuit breakers. In addition, original designs may not have included special attention to the critical areas of production. With these thoughts in mind, the following information should be added to the one-line diagram:

(1) The original system should be identified. The exact nature of the new loads and their approximate locations should be noted.

(2) Critical areas of the system should be highlighted.

(3) The component reliability numbers from Chapter 3 should be inserted so that the reliability performance of the plant can be analyzed on an "if new" basis. (It is preferable to use numbers indigenous to a particular plant whenever this information is available.)

The above information may be too voluminous for clear representation on a single drawing. It may, therefore, be advantageous to include the incoming supply and main feeder circuit breakers (at least) and even major equipment (very large motors or groups requiring the entire capacity of a main feeder position) on one diagram. The load end of the feeders can be detailed on one or more subsequent drawings. After completion of the one-line diagrams, a comprehensive analysis can begin. However, the general inspection covered in 4.4 can, and should, be made concurrent with the preparation of a plant one-line diagram.

The one-line diagram is a picture of an ever changing electric system. The efforts in preparing the diagram and analyzing the system should, therefore, be augmented by a means to capture new pictures of the system (or of proposed systems) as changes are made (or proposed). Therefore, a procedure should be formalized to ensure that all proposals undergo reliability scrutiny (as well as one-line diagram update), and that their effect on the total system is analyzed before the proposal is approved. This process not only maintains the integrity, but it minimizes expense by more effective utilization of existing electrical facilities.

4.4 Plant Reliability Analysis

An inspection analysis of the physical condition of a plant's distribution system can be utilized (hopefully on a continual basis) to improve plant reliability. The following inspection requires little, if any, capital investment while providing a favorable increase in reliability:

(1) Equipment should be periodically checked for proper condition, and programs should be initiated for preventive maintenance procedures as required. (See Chapter 5 for further information.)

(a) Oil in transformers and circuit breakers should be periodically checked for mineral, carbon, and water content as well as level and temperature.

(b) Molded case circuit breakers should be exercised periodically (that is, operated "on" to "off" to "on").

(c) Terminals should be tightened. Each terminal should be inspected for discoloration (overheating) which is generally caused by either a bad connection or equipment overload. Cabinets, etc., should be checked for excessive warmth. Remember that circuit breakers and fuses interrupt as a result of heat in the overload mode.

(d) Surge arresters should be checked for their readiness to operate.

(2) Distribution centers should be checked to see that spare fuses are available. Spare circuit breakers may also be necessary for odd sizes or special applications.

(3) Switches, disconnect switches, bus work, grounds should be checked for corrosion, and unintentional entry of water or corrosive foreign material. It may be wise to operate suspected switches to see that their mechanisms are free, so that faults can be properly isolated and switches safely refused.

(4) The mechanical part of the electrical system should be checked.

(a) The conduit, duct, cable tray, and busway systems should be well supported mechanically, and the grounding system should be electrically continuous. Employees can be shocked or injured if a circuit faults "to ground" without a solid continuous return path to the source interrupter. Supports such as wood poles should be checked for excessive rusting or rotting which would significantly reduce their mechanical strength.

(b) Open-wire circuits should be checked for insulator and surge arrester failure and contamination.

(c) The system's key locations (of open area distribution centers and lines) should be checked for foreign growth such as trees, weeds, shrubs, etc, as well as for general accessibility. The distribution centers should be free from storage of trash, flammables, or even general plant inventory.

(d) Permanent and portable wiring should be checked for fraying or other loss of insulating value.

(e) In general, the system should be checked for any obvious situations where accidents could precipitate an interruption.

(5) The electrical supply room(s) should be thoroughly checked.

(a) The relay and control power fuses should be intact (not blown).

(b) Indicating lights should all be operable and clearly visible.

(c) All targets should be reset so that none show a tripping. Counters (if any) should be checked and the count (number) should be recorded.

(d) The control power, batteries, emergency lighting and emergency generation should be tested and checked to see that they are operational. In many cases, plants have been unable to transfer to their spare circuit or start their standby generator because of dead batteries.

(6) Switches, conduits, busways and duct systems should be checked for overheating. This could be caused by overloaded equipment, severely unbalanced loads or poor connections.

4.5 Circuit Analysis and Action

The first subsequent investigation, following completion of the plant one-line diagram is the analysis of the system to pinpoint design problems. Key critical or vulnerable areas, and overdutied or improperly protected equipment can be located by the following procedure:

(1) Assign faults to various points in the system and note their effect on the system. For example, assume that the cable supply to the air conditioning compressor failed. How long could operations continue? Is any production cooling involved? Are any computer rooms cooled by this system? What would happen if a short circuit (or ground fault) occurred on the secondary terminals of a unit substation? Consideration should be given to relay action (including back-up protection), service restoration procedures, etc, in this "what if" analysis. This review could be called a failure mode and effects analysis.

(2) Calculate feeder loads to verify that all equipment is operating within its rating (do not forget current transformers and other auxiliary equipment). Graphic or demand ammeters (as required) should be used to gather up-to-date information. Fault duties should also be considered. (see Chapter 5 IEEE Std 141–1976, Recommended Practice for Electric Power Distribution for Industrial Plants.

(3) Perform a relay coordination analysis (See IEEE Std 242–1975 Recommended Practice for Protection and Coordination of Industrial and Commercial Power Systems or Chapter 4 of IEEE Standard 141-1976).

(a) Are the relays and fuses properly set or rated for the current load levels?

(b) Is there any new load that has refused critical circuit reliability (or increased vulnerability)?

Obviously, overloaded equipment should be replaced or load transferred so that the equipment can be operated well within its rating. The major projection points—outside the critical areas—should be capable of keeping the system intact by clearing faults and allowing the critical process to continue. The probability of jeopardizing the critical circuits by extraneous electric faults should be minimized, either by physically isolating the critical circuits or by judicial use and proper maintenance of protective devices to electrically sever and isolate faults from critical circuits.

With isolation criteria secure, the investigation should move to the critical circuits themselves to see that proper backup equipment is available and that restoration procedures are adequate. For example, a conveyer system with large rollers may have one motor for each roller, or several hundred motors. The failure rate is 0.0109 per unit year for the motors, or 2 motor failures can be expected annually for a plant with 200 motors. The typical downtime is 65 hours (but could be less for this specific example). In this case, there should be a means of separating the motor from the systems and allowing the conveyer system to continue operation (possibly allowing the roller to idle until the end of a shift), and several spare motors should be available to minimize downtime.

Most plants have a sufficient number of motors to result in several failures per year. The large variety usually precludes the maintenance of a spare motor stock (although their availability can be checked with local distributors). Highly critical nonstandard equipment may require spares. However, each component of the electric system should be viewed in its relationship to the critical process and downtime. (Relay or fuse coordination again plays an important role here.)

The worth of carrying spare parts should be carefully weighed when long process interruptions could result from a single component failure.

4.6 Other Vulnerable Areas

In many plants, the major process is controlled by a small component. This component may be a rectifier system, a computer, or a retrofitted magnetic or punched tape system. The continuity of the electric feed to this controller is just as important to the process as the main machine itself. By proper application of power sources within the device (usually large banks of capacitors) or external uninterruptible power sources, the control can cause the equipment to go into a "safe-hold" position if the power source is interrupted. This continuity (availability) is important to note when thousands of dollars worth of products are being machined in one operation (such as in the aircraft industry). The accuracy and efficacy of a computer or a computer based process is directly related to the "quality" of its environment. This quality is determined by more than just the continuity of the electrical supply. Voltage dips, line noise, ineffective grounding, extraneous electrical and magnetic fields, temperature changes, and even excessively high humidity can adversely affect the accuracy of a computer (or to a lesser extent, a microprocessor). To minimize the probability of errors, the computer should be properly shielded and grounded. It may even be beneficial to install a continuous uninterruptible power supply or transient suppressor equipment on computer circuits where the controlled process is critical.

Testing facilities should have a backup power supply where interruptions could abort long-term testing (that is, tests that span large periods of time). It is important to note that only sufficient power need be supplied to operate the test itself.

Another area of importance is the lighting required for safe operation of the machines. A failure in a particular lighting circuit may reduce the area lighting to a level below that which is necessary to maintain a safe watch over production. Two means of overcoming this vulnerability are (1) emergency task lighting and (2) sufficient lighting such that a single circuit outage does not reduce lighting to an unacceptable level. Another important lighting consideration is the fact that some metal halide lights (HID) require as long as 15 minutes to restart after being extinguished. Since even severe voltage dips can extinguish this type of lighting (a dip that may go virtually unnoticed by production equipment), supplementary lighting is necessary when the metal halide light is a primary source of illumination. Other new high output lamps will restart in 1 to 6 minutes, but this too can cause production problems.

Air, oil, and water systems are frequently important auxiliary inputs upon which production depends. A compressor outage can, for example, cause significant production loss. While failures in these systems are usually mechanical in nature, electrical failures are not uncommon. Pumps are often integral parts of the cooling system in large transformers or even in rectifier circuits, and loss of coolant circulation could either shut down the equipment or significantly reduce production output. Therefore, pumps should be well maintained (mechanically and electrically) when they comprise a significant part of the system, and spare parts may be a wise investment. Ventilation can also be critical to cooling, and ventilator fans are often neglected—until they fail. Hence, periodic maintenance and/or spare ventilator motors may be a good investment.

Some plants rely on a single cable to supply their entire electrical requirements, and many plants rely on single cables for major blocks of load. In these cases, it may be prudent to take several precautionary steps. One possible step would be the periodic testing of cables [2]. Another measure would be the use of spare cables or the storage of a single "portable cable" with permanently made ends (and provisions for installing the portable cable at the various cable terminations in the plant distribution system). Lastly, *advance* (documented) arrangements could be made with a local contractor or the local utility for use of their portable cables (and/or services) on an emergency basis.

Premature equipment failure can result from electric potential that is either too high, too low, excessively harmonic laden, or unbalanced (and also a combination of any or all of these). Voltage tolerances are fairly well established by NEMA and ANSI. However, in [3] a means is provided to evaluate a situation where more than one area deviates from rating. It must be noted that some situations are offsetting such as a high voltage (less than 10 percent high) and unbalanced voltage.

It is important to record and log voltage levels (of all three phases) at various strategic points on a periodic basis (that is, annually) and to occasionally determine the harmonic content in the plant's distribution system. The widespread use of solid-state switching devices has caused an increase in harmonic content in the plant power, but it has been unofficially reported that such devices must approach 50 percent of the plant load before significantly detrimental effects occur. However, the engineer must look at harmonic content in conjunction with other criteria to determine whether there is cause for a significant loss of life in his equipment. Filter circuits are generally used to remove harmful harmonics, and their nature is beyond the scope of this publication. Fluorescent lighting also produces harmonics, but these harmonics are "blocked" by the use of delta–wye transformers.

4.7 Conclusion

The plant engineer should analyze his system electrically and physically and inquire about the city's system. In this analysis, the engineer should:

(1) See that faults are properly isolated and that critical loads are not vulnerable to interruption or delayed repair

(2) Analyze the critical areas and evaluate the need for special restoration equipment, spare parts or procedures

(3) Based on probability and economic analysis, make capital or preventive maintenance investments as indicated by the analysis

(4) Make carefully documented contingency (catastrophy) plans

(5) Check the quality of the power supply from the utility and throughout the plant to determine if the equipment is vulnerable to premature failure

(6) Develop preventive maintenance, checking and logging procedures to

ensure continuous optimum reliability performance of the plant

4.8 References[2]

[1] IEEE COMMITTEE REPORT. Report on Reliability Survey of Industrial Plants, Parts 1-6, *IEEE Transactions on Industry Applications,* vol IA-10, Mar/Apr, July/Aug, Sept/Oct 1974, pp 213–252, 456–476, 681.

[2] LEE, R. New Developments in Cable System Testing, *IEEE Transactions on Industry Applications,* vol IA-13, May/June 1977.

[3] LINDERS, J. R. Effects of Power Supply Variations on AC Motor Characteristics, *IEEE Transactions on Industry Applications,* vol IA-8, July/Aug 1972, pp 383-400.

[4] ANSI/IEEE Std 315-1975, Graphic Symbols for Electrical and Electronics Diagrams.

[5] IEEE Std 141-1976, Recommended Practice for Electric Power Distribution for Industrial Plants.

[6] IEEE Std 242-1975, Recommended Practice for Protection and Coordination of Industrial and Commercial Buildings.

[2]Reference [1] is reprinted in Appendixes A and B.

5. Electrical Preventive Maintenance

5.1 Introduction

The objective of this chapter is to call to your attention the "why" of electrical preventive maintenance and the role it can play in the reliability of distribution systems for industrial plants and commercial buildings. Details of "when" and "how" can be obtained from other sources [1]–[9].[1]

Of the many factors involved in reliability, electrical preventive maintenance usually receives meager emphasis in the design phase and operation of electric distribution systems when in fact it can be a key factor to high reliability. Large expenditures for electric systems are made to provide the desired reliability; however, failure to provide timely and high-quality preventive maintenance leads to system or component malfunction or failure and prevents obtaining the intended design goal.

5.2 Definitions

Electrical preventive maintenance is a system of planned inspection, testing, cleaning, drying, monitoring, adjusting, corrective modification, and minor repair of electrical equipment to minimize or forestall future equipment operating problems or failures; which, depending on equipment type, may require exercising or proof testing.

Electrical equipment is a general term including materials, fittings, devices, appliances, fixtures, apparatus, machines, etc, used as a part of, or in connection with, an electric installation.

5.3 Relationship of Maintenance Practice and Equipment Failure

The Reliability Subcommittee of IEEE Industrial and Commercial Power Systems Committee published the results of a survey which included the effect of maintenance quality on the reliability of electrical equipment in industrial plants [9]. Each participant in the survey was asked to give his opinion of the maintenance quality in his plant. A major portion of the electrical equipment population covered in the survey had a maintenance quality that was classed as "excellent" or "fair." It is interesting to note that

[1] Numbers in brackets correspond to those in the References at the end of this Section.

Table 21
Number of Failures Versus Maintenance Quality for All Equipment Classes Combined

Maintenance Quality	Number of Failures		Percent of Failures Due to Inadequate Maintenance
	All Causes	Inadequate Maintenance	
Excellent	311	36	11.6%
Fair	853	154	18.1%
Poor	67	22	32.8%
None	238	28	11.8%
Total	1469	240	16.4%

Table 22
Percentage of Failure Caused from Inadequate Maintenance Versus Month Since Maintained

Failure, (months since maintained)	All Electrical Equipment Classes Combined	Circuit Breakers	Motors	Open Wire	Trans-formers
Less than 12 months ago	7.4%	*12.5%	8.8%	*0	*2.9%
12–24 months ago	11.2%	19.2%	8.8%	*22.2%	*2.6%
More than 24 months ago	36.7%	77.8%	44.4%	38.2%	36.4%
Total	16.4%	20.8%	15.8%	30.6%	11.1%

*Small sample size; less than 7 failures caused by inadequate maintenance.

maintenance quality had a significant effect on the percentage of all failures blamed on "inadequate maintenance." As shown in Table 21 of the 1469 failures reported from all causes, inadequate maintenance was blamed for 240, or 16.4 percent of all the failures.

The IEEE data also showed that "months since maintained" is an important parameter when analyzing failure data of electrical equipment. Table 22 shows data of failures caused by inadequate maintenance for circuit breakers, motors, open wire, transformers and all equipment classes combined. The percent of failures blamed on inadequate maintenance shows a close correlation with "failure, months since maintained."

From the IEEE data obtained, it was possible to calculate "failure rate multipliers" for transformers, circuit breakers and motors based upon the "maintenance quality." These "failure rate multipliers" are shown in Table 23 and can be used to adjust the equipment failure

Table 23
Equipment Failure Rate
Multipliers Versus Maintenance Quality

Maintenance Quality	Trans-formers	Circuit Breakers	Motors
Excellent	0.95	0.91	0.89
Fair	1.05	1.06	1.07
Poor	1.51	1.28	1.97
All	1.0	1.0	1.0
Perfect maintenance	0.89	0.79	0.84

rates shown in Chapter 3. "Perfect" maintenance quality has zero failures caused by inadequate maintenance.

5.4 Design for Electrical Preventive Maintenance

Electrical preventive maintenance should be a prime consideration for any new electrical equipment installation. Quality, installation, configuration, and application are fundamental prerequisites in attaining a satisfactory preventive maintenance program. A system that is not adequately engineered, designed, and constructed will not provide reliable service, regardless of how good or how much preventive maintenance is accomplished.

One of the first requirements in establishing a satisfactory and effective preventive maintenance program is to have good-quality electrical equipment which is properly installed. Examples of this are as follows:

(1) Large exterior bolted covers on switchgear or large motor terminal compartments are not conducive to routine electrical preventive maintenance inspections, cleaning, and testing. Hinged and gasketed doors with a three point locking system would be much more satisfactory.

(2) Space heater installation in switchgear or an electric motor is a vital necessity in high humidity areas. This reduces condensation on critical insulation components. Installation of ammeters in the heater circuit is an added tool for operating or maintenance personnel to monitor their operation.

(3) Motor insulation temperatures can be monitored by use of resistance temperature detectors which provide an alarm indication at a selected temperature depending on the insulation class. Such monitoring indicates that the motor is dirty and/or air passages are plugged.

The distribution system configuration and features should be such that maintenance work is permitted without load interruption or with only minimal loss of availability. Often, equipment preventive maintenance is not done or is deferred because load interruption is required to a critical load or to a portion of the distribution system. This may require the installation of alternate electrical equipment and circuits to permit routine or emergency maintenance on one circuit while the other one supplies the critical load that cannot be shutdown.

Electrical equipment which is improperly applied will not give reliable service regardless of how good or how much preventive maintenance is accomplished. The most reasonably accepted measure is to make a corrective modification.

5.5 Electrical Equipment Preventive Maintenance

Electrical equipment deterioration is normal. Unchecked, the deterioration can progress and cause malfunction or an electrical failure. Electrical equipment preventive maintenance procedures should be developed to accomplish four basic functions, that is, keep it clean, dry, sealed tight, and minimize the friction. Water, dust, high or low ambient temperature, high humidity, vibration, component quality, and countless other conditions can affect proper operation of electrical equipment. Without an effective electrical preventive maintenance program the risk of a serious electrical failure increases.

A common cause of electrical failure is dust and dirt accumulation and the presence of moisture. This can be in the form of lint, chemical dust, day-to-day accumulation of oil mist and dirt particles, etc. These deposits on the insulation, combined with oil and moisture, become conductors and are responsible for tracking and flashovers. Deposits of dirt can cause excessive heating and wear, and decrease apparatus life. Electrical apparatus should be operated in a dry atmosphere for best results; but this is often impossible, so precautions should be established to minimize entrance of moisture. Moisture condensation in electrical apparatus can cause copper or aluminum oxidation and connection failure.

Loose connections are another cause of electrical failures. Electrical connections should be kept tight and dry. Creep or cold flow is a major cause of joint failure. Mounting hardware and other bolted parts should be checked during routine electrical equipment servicing.

Friction can affect the freedom of movement of electrical devices and can result in serious failure or difficulty. Dirt on moving parts can cause sluggishness and improper electrical equipment operations such as arcing and burning. Checking the mechanical operation of devices and manually or electrically operating any device that seldom operates should be standard practice.

Procedures and practices should be initiated to substantiate that electrical equipment is kept clean, dry, sealed tight, and with minimal friction by visual inspection, exercising, and proof testing. Electrical preventive maintenance should be accomplished on a regularly scheduled basis as determined by inspection experience and analysis of any failures that occur.

An electrical preventive maintenance program certainly will not eliminate all failures, but it will minimize their occurrence. Some of the key elements in establishing a program are as follows:

(1) Establish an "Equipment Service Library" consisting of bulletins, manuals, schematics, parts lists, failure analysis reports, etc. The bulletins and manuals are normally provided by the electrical equipment manufacturer. Often they are not taken very seriously after equipment installation and are lost, misplaced, or discarded, but this documentation is vital to develop electrical preventive maintenance procedures and to aid in training.

(2) In addition to the above documentation, each in-service failure should be thoroughly investigated and the cause determined and documented. Generally, it will be found that timely and adequate electrical preventive maintenance could have prevented the failure. If correctable by electrical preventive maintenance the corrective action should be included on the work list. If the failure was caused by a weak component, then all like equipment should be modified as soon as possible.

"Failure analysis" plays a major part in an electrical preventive maintenance program.

(3) Provide the training necessary to accomplish the program that has been established. The techniques utilized in performance of an electrical preventive maintenance program are extremely important. The success or failure relies on the qualification and know-how of personnel performing the work; therefore, training in electrical preventive maintenance techniques is a major objective. Servicing of electrical equipment requires better than average skills and special training. Properly trained and adequately equipped maintenance personnel must have a very thorough knowledge of the equipment operation. It must be able to make a thorough inspection and also accomplish repairs. For example, special training in the use of the dc high-potential dielectric tests or megger tests as well as the interpretation of the results may be required.

(4) A good record system should be developed which will show the repairs required by equipment over a long period of time. On each regular inspection, variations from normal conditions should be noted. The frequency and magnitude of the work should then be increased or decreased according to an analysis of the data. Avoid performing too much maintenance work as this can contribute to failures. The records should reflect availability of spare parts, service attitude of equipment manufacturers, major equipment failures to date and time required for repairs, etc. These records are not only useful in planning and scheduling electrical preventive maintenance work, but are useful in evaluating equipment performance for future purchases.

5.6 References

[1] ANSI/NFPA 70-1978, National Electrical Code[2,3]

[2] HUBERT, C. I. Preventive Maintenance of Electrical Equipment, McGraw-Hill Book Co.

[3] Maintenance Hints, Westinghouse Electric Corp., Pittsburgh, PA.

[4] SHAW, E. T. Inspection and Test of Electrical Equipment, Westinghouse Electric Division, Pittsburgh, PA.

[5] SMEATON, R. W. Motor Application and Maintenance Handbook. McGraw Hill Book Co, New York, NY 1969.

[6] Factory Mutual Systems Transformer Bulletin 14-8, October 1976. Public Information Division, 1151 Boston-Providence Turnpike, Norwood, MA

[7] MILLER, H. N. DC Hypot Testing of Cables, Transformers and Rotating Machinery, Manual P-16086. Associated Research Inc., Chicago, IL

[8] CURDTS, E. B. Insulation Testing by D-C Methods, Technical Publication 22T1-1971 James G. Biddle Co, Plymouth Meeting, p 2.

[9] IEEE COMMITTEE REPORT. Report on Reliability Survey of Industrial Plants, Part 6, *IEEE Transactions on Industry Applications*, vol IA-10, July/August and September/October 1974, pp 456–476, 681.[4]

---

[2]American National Standards Institute, 1430 Broadway, New York, NY 10018.

[3]National Fire Protection Association, 470 Atlantic Avenue, Boston, MA 02210.

[4]Reference [9] is reprinted in Appendix B.

[10] *ANSI/IEEE C57.106-1977,* IEEE Guide for Acceptance and Maintenance of Insulating Oil in Equipment

[11] *ANSI/IEEE Std 43-1974,* IEEE Recommended Practice for Testing Insulation Resistance of Rotating Machinery

[12] *ANSI/IEEE Std 56-1977,* IEEE Guide for Insulation Manitenance of Large AC Rotating Machinery (10 000 kVA and Larger)

[13] *ANSI/IEEE Std 95-1977,* IEEE Recommended Practice for Insulation Testing of Large AC Rotating Machinery with High Direct Voltage

[14] *IEEE Std 450-1980,* IEEE Recommended Practice for Maintenance, Testing, and Replacement of Large Lead Storage Batteries for Generating Stations and Substations

[15] *IEEE Std 62-1978,* IEEE Guide for Field Testing Power Apparatus Insulation

6. Emergency and Standby Power

6.1 Introduction

When a reliability analysis has been completed, the rate at which power failures occur and the expected duration of those power failures can be predicted at most points of utilization equipment. This knowledge can be used to determine whether there is a need to increase the reliability or availability of delivered power supplied to particular utilization points. Emergency or standby power can readily be used to improve both reliability and availability of delivered power. A cost-reliability tradeoff decision must be made to improve reliability or availability of power only to those areas which can justify the cost for such improvement. Various types of emergency or standby systems are ideally suited to providing large improvements to relatively small sections of a power system.

6.2 Reliability and Availability

An evaluation of each piece of utilization equipment must be made to determine actual needs. The difference between reliability and availability of supplied power must be clearly understood. Reliability is an indication of how many power failures can be expected over a certain period of time, while availability is an indication of total downtime due to lack of power over a certain period. Both reliability and availability requirements for control power for a boiler would certainly be greater than those for a room air conditioner.

Many power-consuming operations require a high degree of power-supply reliability with little concern for availability. A power failure during the vulcanizing cycle of a rubber manufacturing process will cause loss of steam and errors in the time/temperature control for proper curing. This results in the product being scrapped. The difference in loss between a power failure of one-minute duration and one of 30-minute duration is minimal. Thus a power system that experiences two power failures of 30 minutes each is more desirable than a system that experiences six power failures of one minute each, even though the downtime is 10 times greater in the first case. Other power utilizing equipment demands a high degree of power supply availability with little concern for reliability. A power failure to a process which stamps out metal parts will cause little loss due to the power failure itself; but there will be a loss directly related to the length of the power failure. Thus, a power system that experiences six power failures of one minute each is more suitable than one that experiences two power failures of 30 minutes each, even though the first case has a failure rate three times higher and, thus, a lower reliability.

6.3 System Selection

The type of emergency or standby power system to use depends on what the system is expected to accomplish. Can the equipment or process tolerate a power failure of one millisecond, 10 seconds, or of one minute? For how long a period of time does the emergency or standby power system have to perform its intended function? For hours, minutes, or seconds?

An off-line system is one that is dormant until called upon to operate, such as a diesel generator which is started up when a power failure occurs. An on-line system is one that is operating at all times, such as an inverter supplied by dc power via the primary power source through a battery charger. The above system utilizes batteries on float charge to supply the inverter if a power failure strikes the primary power source.

The selection of an off-line engine-driven generator for the rubber manufacturing process mentioned above would be a misapplication. The off-line system can improve the availability of delivered power, but cannot improve the reliability. The transfer device or devices have a failure rate of their own and, thus, actually reduce the reliability of delivered power. When the primary power source is being utilized, a failure of transfer device may cause loss of power which would not have occurred if the off-line system had not been installed. The selection of an off-line system for the metal stamping process is a proper application. The off-line system can reduce downtime, resulting in higher availability.

A study must be undertaken to determine the systems capable of performing the desired function. Systems are available to provide reliable power to overcome the problems encountered due to power failures ranging from milliseconds to many hours. More than one type of system may be suitable for a particular application. Selection of the proper system will then depend on first cost, operating and owning cost such as maintenance and fuel requirements, system reliability, output power quality, expansion capacity, and environmental considerations.

6.4 Descriptions and Applications of Available Systems

6.4.1 *General.* The following information contains data on some commonly used systems. While power sources such as solar and chemical may become viable in the future, they are not in common use and will not be discussed here.

6.4.2 *Engine Driven Generators.* These units are available in sizes from 1 kW to several thousand kW. Fuels commonly used are diesel, gasoline, and natural or liquefied petroleum (LP) gas. If kept warm they will dependably come on line in 8 to 15 seconds. Except for small units, installed costs range from \$250 to \$400 per kW. Diesel units are generally heavier duty, have less costly fuel, and fire danger is lower than for gasoline units. Gasoline-driven units range up to 100 kW and have a lower initial cost than diesel sets. Natural and LP gas engines provide quick starting after long shutdown periods because of the inherently fresh fuel. One of the drawbacks may be the lack of assurance of fuel supplies when the system is needed. Engine-driven generators are generally applied as off-line units for reducing downtime or in combination with a mechanical stored energy system or a small uninterruptible power supply to improve both reliability and availability of delivered power.

6.4.3 *Turbine-Driven Generators..* Two types of turbines can be used for prime movers, steam or gas. Since steam is generally not available when a power failure has occurred, only the gas prime mover will be discussed. Installed costs for medium-sized units range from $200 to $400 per kW.

Gas turbines can utilize various grades of oil as well as natural and propane gas. Sizes generally range from 100 kW to several thousand kW. Gas turbine generators can be placed on line in 20 seconds for smaller units up to several minutes for larger units. They can more easily be rooftop mounted since their physical size and weight per kilowatt are less than for engine-driven units. Turbine driven generators are almost exclusively applied as off-line systems.

6.4.4 *Mechanical Stored Systems.* This type of system is comprised of a rotating flywheel which converts its rotating kinetic energy into electric power. It is generally applied as an on-line system. Depending on frequency requirements of the load, a typical mechanical stored energy system can ride through a power failure for up to two seconds. Thus, its main use is as a buffer to mechanically filter out transients. See Fig 5. A supply time of 15 seconds can be attained by using an eddy current clutch and driving the flywheel at a higher speed than the generator it operates. This type of system may allow an engine driven prime mover to come up to speed, either to drive a separate generator or to maintain the speed of the flywheel and its associated generator. See Fig 6. Such systems have a history of maintenance problems. There are several other hybrid systems which utilize dc drive motors, batteries, engines, and turbines. System costs range from $200 to $1000 per kW.

6.4.5 *Inverter/Battery Systems.* A simple off-line inverter system is shown in

Fig 5
Simple Inertia Driven
"Ride Through" System

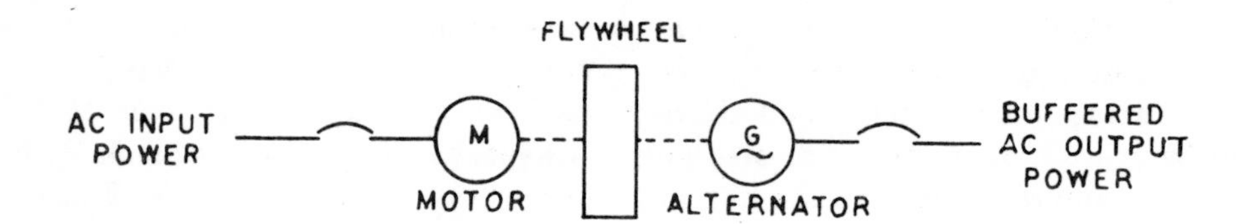

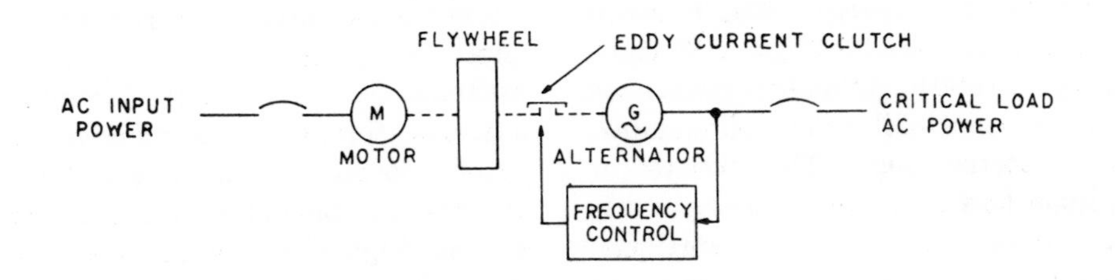

Fig 6
Constant Frequency Inertia System

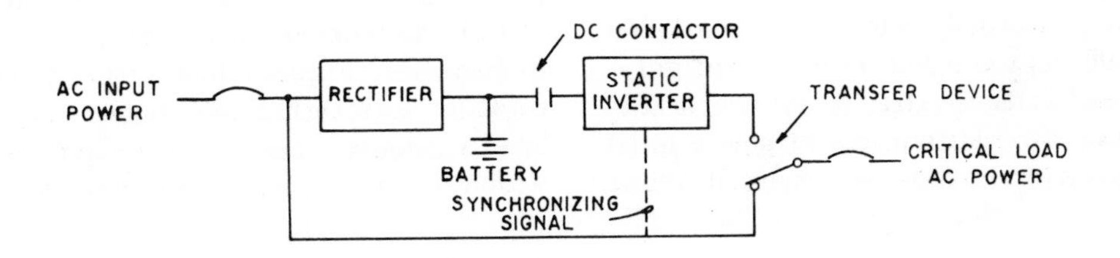

Fig 7
Short Interruption Static Inverter System

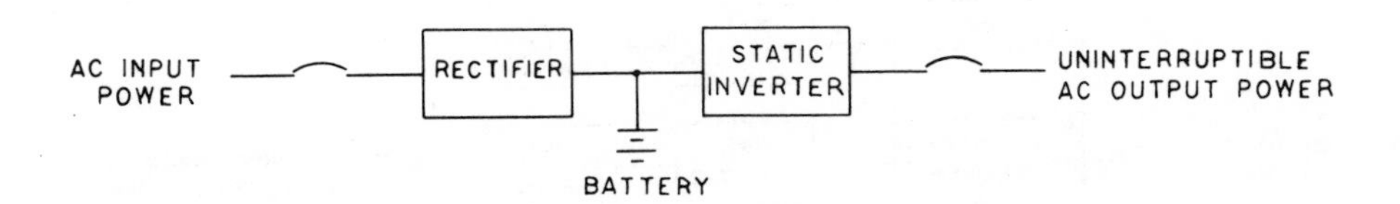

Fig 8
Non-Redundant Uninterruptible Power Supply

Fig 7. The system is not an uninterruptible power supply. The transfer time for a mechanical transfer will cause a power interruption of 60 to 190 milliseconds. A static transfer switch is more costly, but will result in a much shorter interruption. The contactor closes upon loss of primary power and is in the circuit to prevent continual energization of the static inverter whose efficiency is approximately 70 percent and, thus, wastes energy while energized.

Figure 8 shows the most widely used system for supplying uninterruptible power. The load is basically free of power interruptions, transient disturbances, and voltage and frequency variations. Installed costs range from $200 to $800 depending on system configuration and battery size. A failure of the inverter will cause a loss of power until the inverter is repaired or until prime power can be connected directly to the load.

Figure 9 shows a redundant uninterruptible power supply with static switches to clear a faulted inverter. The batteries for this system are required to supply power only until the diesel generators can be placed on line. The system in Figure 9 is much more reliable than that shown in Figure 8, but is more expensive. Depending on amount of redundancy, auxiliary equipment, and required battery size, the cost per kW ranges from $1000 to $3000. Installation requirements can be impressive for the battery. A battery sized to provide power for a 250 kW inverter for one hour will weigh approximately 25 tons.

6.4.6 *Mechanical Uninterruptible Power Supplies.* Figure 10 shows a typical rotating uninterruptible power supply. The ac motor drives the dc generator which in turn supplies power for the dc

Fig 9
Redundant Uninterruptible Power Supply

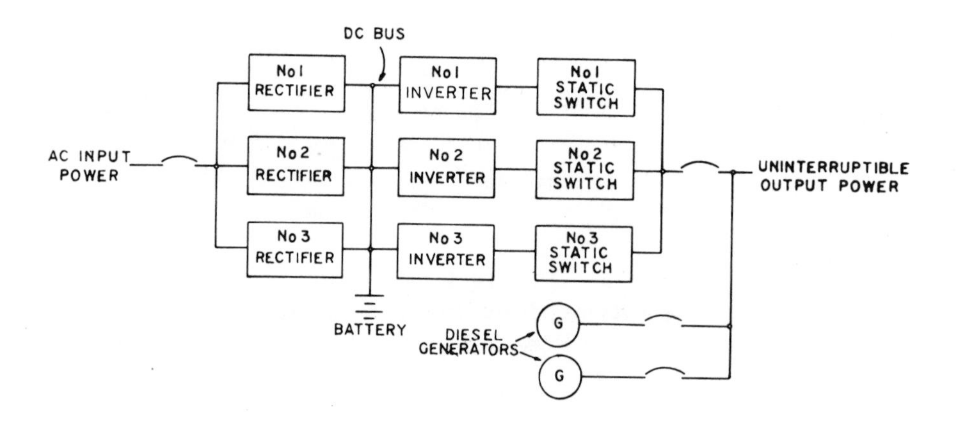

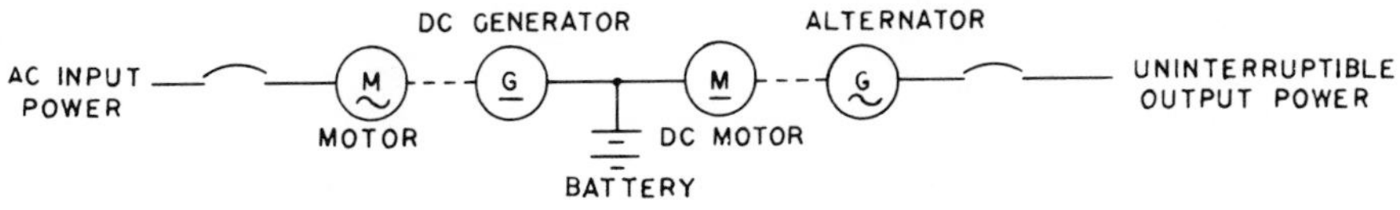

Fig 10
Rotating Uninterruptible Power Supply

motor which drives the ac generator. The battery will provide power for the dc motor upon loss of primary power. The ac generator provides uninterruptible power to the load. The lack of moving parts in the static inverters and rectifiers has proven to be a strong selling point over the mechanical uninterruptible power supplies.

6.5 Selection and Application Data. The figures and system descriptions presented here are only a few of the many types of system and hybrid systems available. For comprehensive selection and application data, reference is made to IEEE Std 446–1980, Recommended Practice for Emergency and Standby Power Systems.

7. Examples of Reliability Analysis and Cost Evaluation

7.1 Examples of Reliability and Availability Analysis of Common Low-Voltage Industrial Power Distribution Systems

7.1.1 *Quantitative Reliability and Availability Predictions.* A description is given of how to make quantitative reliability and availability predictions for proposed new configurations of industrial power distribution systems. Six examples are worked out, including a simple radial system, a primary-selective system, and a secondary-selective system. A brief tabulation is also given of pertinent reliability data needed in order to make the reliability and availability predictions. The simple radial system analyzed had an average number of forced hours of downtime per year that was 19 times larger than a secondary-selective system; the failure rate was six times larger. The importance of two separate power supply sources from the electric utility has been identified and analyzed. This approach could be used to assist in cost–reliability tradeoff decisions in the design of the power distribution system.

7.1.2 *Introduction.* An industrial power distribution system may receive power at 13.8 kV from an electric utility and then distribute the power throughout the plant for use at the various locations. One of the questions often raised during the design of the power distribution system is whether there is a way of making a quantitative comparison of the failure rate and the forced hours downtime per year of a secondary-selective system with a primary-selective system and a simple radial system. This comparison could be used in cost-reliability and cost-availability tradeoff decisions in the design of the power distribution system. The estimated cost of power outages at the various plant locations could be factored into the decision as to which type of power distribution system to use. The decisions could be based upon "total owning cost over the useful life of the equipment" rather than "first cost."

Six examples of common low-voltage industrial power distribution systems are analyzed in this section:

(1) Example 1—Simple radial

(2) Example 2—Primary selective to 13.8 kV utility supply

(3) Example 3—Primary selective to load side of 13.8 kV circuit breaker

(4) Example 4—Primary selective to Primary of transformer

(5) Example 5—Secondary selective

(6) Example 6—Simple radial with spares

Only forced outages of the electrical equipment are considered in the six examples. It is assumed that scheduled maintenance will be performed at times when 480 V power output is not needed. The frequency of scheduled outages and the average duration can be estimated, and, if necessary, these can be added to the forced outages given in the six examples.

When making a reliability study, it is necessary to define what is a failure of the 480 volt power. Some of the failure definitions for 480 volt power that are often used are as follows:

(1) Complete loss of incoming power for more than 1 cycle

(2) Complete loss of incoming power for more than 10 cycles

(3) Complete loss of incoming power for more than 5 seconds

(4) Complete loss of incoming power for more than 2 minutes

Definition (3) will be used in the six examples given. This definition of failure can have an effect in determining the necessary speed of automatic throwover equipment that is used in primary-selective or secondary-selective systems. In some cases when making reliability studies, it might be necessary to further define what is "complete loss of incoming power"; for example, "voltage drops below 70 percent."

One of the main benefits of a reliability and availability analysis is that a disciplined look is taken at the alternative choices in the design of the power distribution system. By using published reliability data collected by a technical society from industrial plants, the best possible attempt is made to use historical experience to aid in the design of the new system.

7.1.3 *Definition of Terminology.* The definition of terms is given in Section 2.1.3. The units that are being used for "failure rate" and "average downtime per failure" are:

λ = Failure rate (failures per year)

r = average downtime per failure (hours per failure) = Average time to repair or replace a piece of equipment after a failure. In some cases this is the time to switch to an alternate circuit when one is available.

7.1.4 *Procedure for Reliability and Availability Analysis.* The "minimal cut-set" method for system reliability evaluation is described in Sections 2.1.6, 2.1.8, and 2.1.9. The quantitative reliability indexes that are used in the six examples are the failure rate and the forced hours downtime per year. These are calculated at the 480 volt point of use in each example. The failure rate λ is a measure of unreliability. The product λr,(failure rate × average downtime per failure) is equal to the forced hours downtime per year and can be considered a measure of forced unavailability since a scale factor of 8760 converts one quantity into the other. The average downtime per failure r could be called restorability.

The necessary formulas for calculating the reliability indexes of the

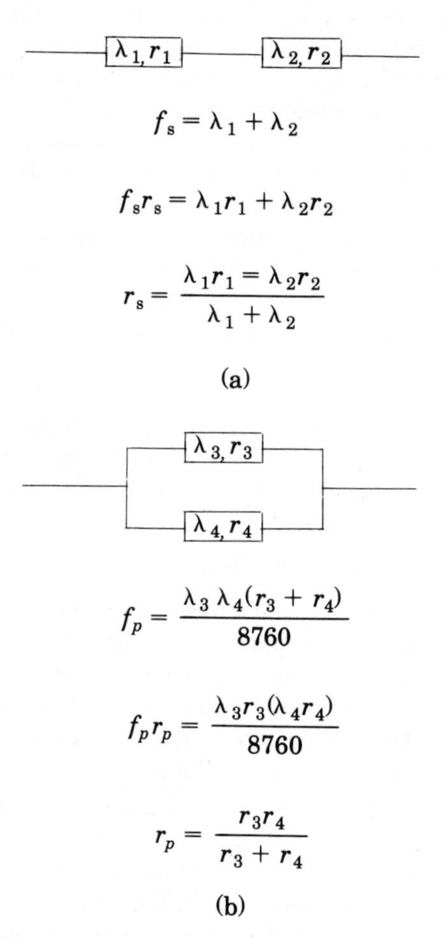

Nomenclature:
f = frequency of failures
λ = failures per year
r = average hours of downtime per failure
s = series
p = parallel

Fig 11
Formulas for Reliability Calculations
(a) Reparable Components in Series (both must work for success)

(b) Reparable Components in Parallel (one or both must work for success)

minimal cut-set approach are given in Table 1. A sample using these formulas is shown in Fig 11 for two components in series and two components in parallel. In these samples the scheduled outages are assumed to be zero and the units for λ and r are, respectively, failures per year and hours downtime per failure. The formulas in both Table 1 and Fig 11 assume the following:

(1) The component failure rate is constant with age

(2) The outage time after a failure has an exponential distribution. (Probability of outage time exceeding τ is $\epsilon^{-\tau/r}$

(3) Each failure event is independent of any other failure event.

(4) The component "up" times are much larger than "down" times:

$$1 - \frac{\lambda_i r_i}{8760} >> \frac{\lambda_i r_i}{8760}$$

The reliability data to be used for the electrical equipment and the electric utility supply are given in 7.1.5.

7.1.5 *Reliability Data from 1973–1975 IEEE Surveys.* In order to make a reliability and availability analysis of a power distribution system, it is necessary to have data on the reliability of each component of electrical equipment used in the system. Ideally these reliability data should come from field use of the same type of equipment under similar environmental conditions and similar stress levels. In addition, there should be a sufficient number of field failures in order to represent an adequate sample size. It is believed that eight field failures are the minimum number necessary in order to have a reasonable chance of determining a failure rate to within a factor of 2. The

Table 24
Reliability Data from IEEE Reliability Survey of Industrial Plants [1]
(See Table 11)

Equipment Category	λ, Failures per Year	r, Hours of Downtime per Failure	$\lambda \cdot r$, Forced Hours of Downtime per Year	Data Source in IEEE Survey [1], Table No.
Protective relays	0.0002	5.0	0.0010	19
Metalclad drawout circuit breakers				
0–600 V	0.0027	4.0	0.0108	5, 50
Above 600 V	0.0036	83.1*	0.2992	5, 51
Above 600 V	0.0036	2.1**	0.0076	5, 51
Power cables (1000 circuit ft)				
0–600 V, above ground	0.00141	10.5	0.0148	13
601–15 000 V, conduit below ground	0.00613	26.5*	0.1624	13, 56
601–15 000 V, conduit below ground	0.00613	19.0**	0.1165	13, 56
Cable terminations				
0–600 V, above ground	0.0001	3.8	0.0004	17
601–15 000 V, conduit below ground	0.0003	25.0	0.0075	17
Disconnect switches enclosed	0.0061	3.6	0.0220	9
Transformers				
601–15 000 V	0.0030	342.0*	1.0260	4, 48
601–15 000 V	0.0030	130.0**	0.3900	4, 48
Switchgear bus—bare				
0–600 V (connected to 7 breakers)	0.0024	24.0	0.0576	10
0–600 V (connected to 5 breakers)	0.0017	24.0	0.0408	10
Switchgear bus—insulated				
601–15 000 V (connected to 1 breaker)	0.0034	26.8	0.0911	10
601–15 000 V (connected to 2 breakers)	0.0068	26.8	0.1822	10

*Repair failed unit.
**Replace with spare.

types of reliability data needed on each component of electrical equipment are:

(1) Failure rate, failures per year
(2) Average downtime to repair or replace a piece of equipment after a failure, hours per failure

These reliability data on each component of electrical equipment can then be used to represent historical experience for use in cost–reliability and cost–availability tradeoff studies in the design of new power distribution systems.

During 1973–1975 the Power Systems Reliability Subcommittee of the Industrial and Commercial Power Systems Committee conducted and published [1] [2] surveys of electrical equipment reliability in industrial plants. See Appendices A, B, and D for the data. See Chapter 3 for a summary of the data.

Table 25
Failure Modes of Circuit Breakers

Percentage of Total Failures in Each Failure Mode

(See Table 14)

Percent of Total Failures (All Voltages)	Failure Characteristic
	Backup protective equipment required
9	Failed while opening
	Other circuit breaker failures
7	Damaged while successfully opening
32	Failed while in service (not while opening or closing)
5	Failed to close when it should
2	Damaged while closing
42	Opened when it shouldn't
1	Failed during testing or maintenance
1	Damage discovered during testing or maintenance
1	Other
100	Total Percentage

Table 26
IEEE Survey of Reliability of Electric Utility Power Supplies to Industrial Plants

(See Table 12)

Number of Circuits (All Voltages)	λ, Failures per Year	r, Hours of Downtime per Failure	$\lambda \cdot r$, Forced Hours of Downtime per Year
Single circuit	1.956	1.32	2.582
Double circuit			
Loss of both circuits*	0.312	0.52	0.1622
Calculated value for loss of Source 1 (while Source 2 is OK)	1.644	0.15*	0.2466

*Data for double circuits had all circuit breakers closed.

**Manual switchover time of 9 min to source 2.

These reliability surveys of electrical equipment and electric utility power supplies were extensive. The pertinent failure rate and average downtime per failure information for the electrical equipment are given in Table 24. In compiling these data, a failure was defined as any trouble with a power system component that causes any of the following effects:

(1) Partial or complete plant shutdown, or below-standard plant operation
(2) Unacceptable performance of user's equipment
(3) Operation of the electrical protective relaying or emergency operation of the plant electric system
(4) Deenergization of any electric circuit or equipment

A failure of an in-plant component causes a forced outage of the component, and the component thereby is unable to perform its intended function until it is repaired or replaced.

In addition to the reliability data for electrical equipment shown in Table 24, there are some "failure modes" of circuit breakers that require backup protective equipment to operate; for example, "failed to trip" or "failed to interrupt." Both of these failure modes would require that a circuit breaker farther up the line be opened, and this would result in a larger part of the power distribution system being disconnected. Reliability data on the "failure modes of circuit breakers" are shown in Table 25. These data are used for the 480 volt circuit breakers in all six examples discussed in this section. It will be assumed that the "flashed over while open" failure mode for circuit breakers and disconnect switches has a failure rate of zero.

The failure rate and average downtime per failure data for the electric utility power supplies are given in Table 26. This includes both single circuit and double-circuit reliability data. The two power sources in a double-circuit utility supply are not completely independent, and the reliability and availability analysis must take this into consideration. This subject is discussed further in 7.1.15.

7.1.6 *Example 1—Reliability and Availability Analysis of a Simple Radial System.*

Description of Simple Radial System. A simple radial system is shown in Fig 12. Power is received at 13.8 kV from the electric utility. It then goes through a 13.8 kV circuit breaker inside the industrial plant, 600 feet of cable in underground conduit, an enclosed disconnect switch, to a transformer that reduces the voltage to 480 volts, then through a 480 volt main circuit breaker, a second 480 volt circuit breaker, 300 feet of cable in above ground conduit, to the point where the power is used in the industrial plant.

Results—Simple Radial System. The results from the reliability and availability calculations are given in Table 27. The failure rate and the forced hours downtime per year are calculated at the 480 volt point of use.

The relative ranking of how each component contributes to the failure rate is of considerable interest. This is tabulated in Table 28.

The relative ranking of how each component contributes to the forced hours downtime per year is also of considerable interest. This is given in Table 29.

It might be expected that the power distribution system would be shut down

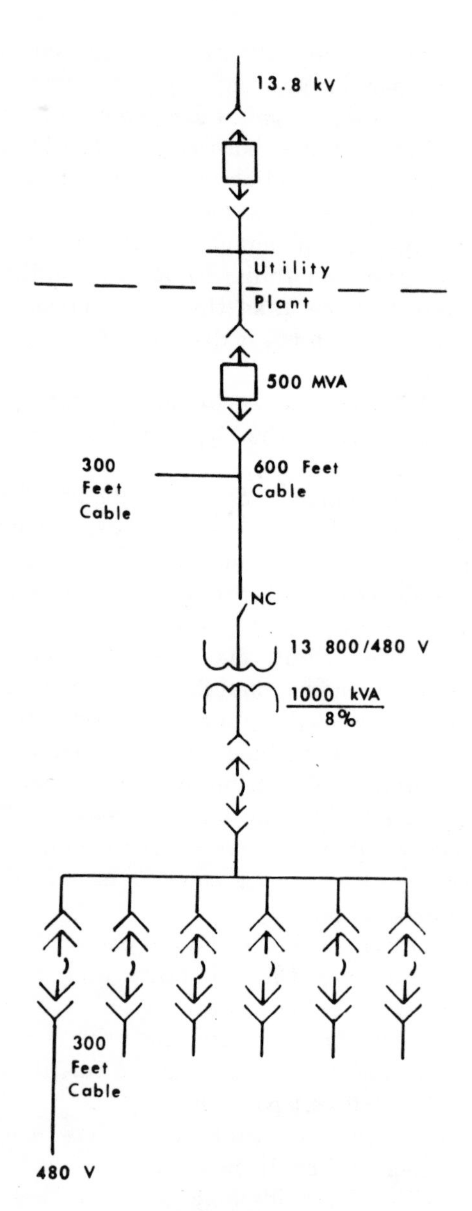

Fig 12
Simple Radial System
Example 1

once every two years for scheduled maintenance for a period of 24 hours. These shutdowns would be in addition to the outage data given in Tables 27, 28, and 29.

Conclusions—Simple Radial System. The electric utility supply is the largest contributor to both the failure rate and the forced hours downtime per year at the 480 volt point of use. A significant improvement can be made in both the failure rate and the forced hours downtime per year by having two sources of power at 13.8 kV from the electric utility. The improvements that can be obtained are shown in Examples 2, 3 and 4 using "Primary-Selective" and in Example 5 using "Secondary-Selective."

The transformer is the second largest contributor to forced hours downtime per year. The transformer has a very low failure rate, but the long outage time of 342 hours after a failure results in a large forced hours downtime per year. The 13.8 kV circuit breaker is the third largest contributor to forced hours downtime per year, and the fourth largest are the 13.8 kV cables and terminations. This is a result of the average outage time after a failure of 83.1 hours for the 13.8 kV circuit breaker and 26.5 hours for the 13.8 kV cable.

The long outage time after a failure for the transformer, 13.8 kV circuit breaker, and the 13.8 kV cable are all based upon "repair failed unit." These outage times after a failure can be reduced significantly if the "replace with spare" times shown in Table 24 are used instead of "repair failed unit." This is done in Example 6, using a simple radial system with Spares.

7.1.7 *Example 2—Reliability and Availability Analysis of Primary-Selective to 13.8 kV Utility Supply.*

Table 27
Simple Radial System
Reliability and Availability of Power at 480 V—Example 1

Component	λ, Failures per Year	$\lambda \cdot r$, Forced Hours of Downtime per Year
13.8 kV power source from electric utility	1.956	2.582
Protective relays (3)	0.0006	0.0030
13.8 kV metalclad circuit breaker	0.0036	0.2992*
Switchgear bus—insulated (connected to 1 breaker)	0.0034	0.0911
Cable (13.8 kV); 900 ft, conduit below ground	0.0055	0.1458*
Cable terminations (6) at 13.8 kV	0.0018	0.0450
Disconnect switch (enclosed)	0.0061	0.0220
Transformer	0.0030	1.0260*
480 V metalclad circuit breaker	0.0027	0.0108
Switchgear bus—bare (connected to 7 breakers)	0.0024	0.0576
480 V metalclad circuit breaker	0.0027	0.0108
480 V metalclad circuit breakers (5) (failed while opening)	0.0012	0.0048
Cable (480 V); 300 ft conduit above ground	0.0004	0.0044
Cable terminations (2) at 480 V	0.0002	0.0008
Total at 480 V output	1.9896	4.3033

*Data for hours of downtime per failure are based upon *repair failed unit.*

Description — Primary-Selective to 13.8 kV Utility Supply. The primary-selective to 13.8 kV utility supply is shown in Fig 13. It is a simple radial system with the addition of a second 13.8 kV power source from the electric utility; the second power source is normally disconnected. In the event that there is a failure in the first 13.8 kV utility power source, then the second 13.8 kV utility power source is switched on to replace the failed power source. Assume that the two utility power sources are synchronized. *Example 2a*—Assume a 9-minute "manual switchover time" to utility power source No 2 after a failure of source No 1. *Example 2b*—Assume an "automatic switchover time" of less than 5 seconds after a failure is assumed. (loss of 480 volt power for less than 5 seconds is not counted as a failure.)

Results—Primary-Selective to 13.8 kV Utility Supply.

Example 2a—If the time to switch to a second utility power source takes 9 minutes after a failure of the first source, then there would be a power supply failure of 9 minutes duration. Using the data from Table 26, for double-circuit utility supplies, this would occur 1.644 times per year (1.956 − 0.312). This in addition to losing both power sources simultaneously 0.312 times per year for an average outage time of 0.52 hours. If these utility supply data are added together and substituted into Table 27 on the Simple Radial System, it would result in reducing the forced hours downtime per year at the 480 volt point of use from 4.3033 to 2.1291. The failure rate would stay the same at 1.9896 failures per year. These results are given in Table 30.

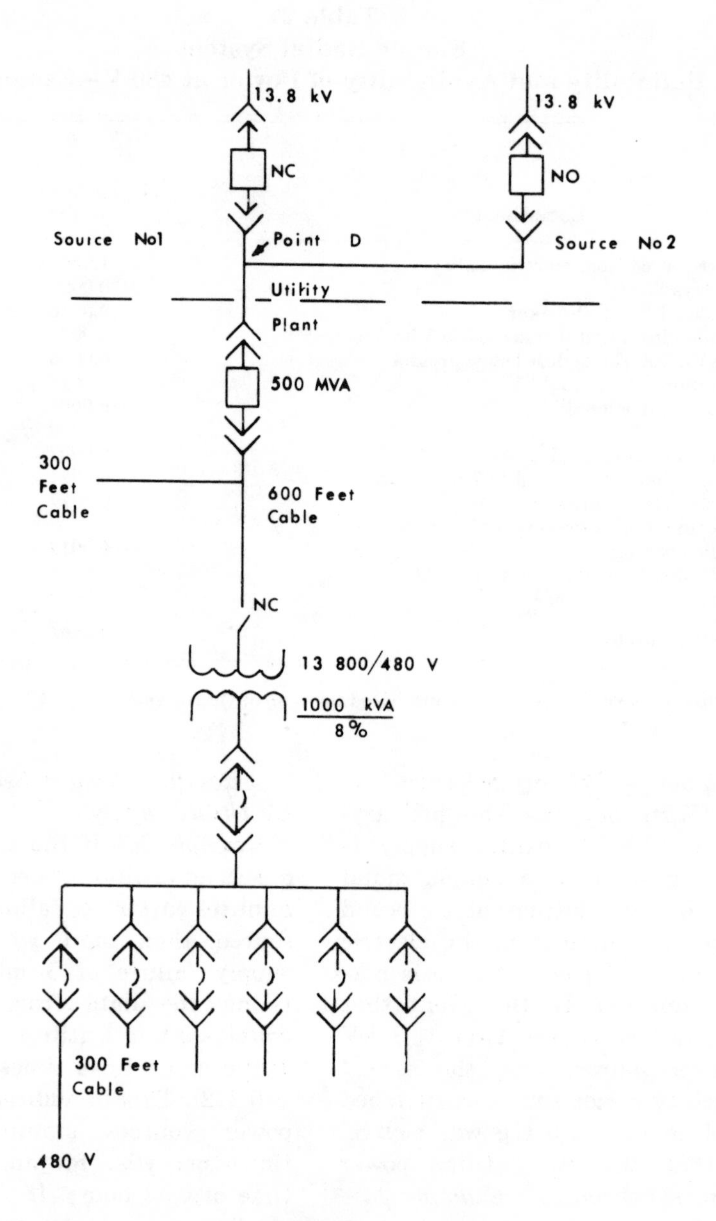

Fig 13
Primary Selective to 13.8 kV
Utility Supply
Example 2

Table 28
Simple Radial System
Relative Ranking of Failure Rates

	λ, Failures per Year
1. Electric utility	1.956
2. 13.8 kV cable and terminations	0.0073
3. Disconnect switch	0.0061
4. 13.8 kV circuit breaker	0.0036
5. Switchgear bus—insulated	0.0034
6. Transformer	0.0030
7. 480 V circuit breaker	0.0027
8. 480 V circuit breaker (main)	0.0027
9. Switchgear bus—bare	0.0024
10. 480 V circuit breakers (5) (failed while opening)	0.0012
11. 480 V cable and terminations	0.0006
12. Protective relays (3)	0.0006
Total	1.9896

Example 2b—If the time to switch to a second utility power source takes less than 5 seconds after a failure of the first source, then there would be no failure of the electric utility power supply. The only time a failure of the utility power source would occur is when both sources fail simultaneously. It will be assumed that the data shown in Table 26 are applicable for loss of both power supply circuits simultaneously. This is 0.312 failures per year with an average outage time of 0.52 hours. If these values of utility supply data are substituted into Table 27, it would result in reducing the forced hours downtime per year from 4.3033 to 1.8835 hours per year at the 480 volt point of use; the failure rate would be reduced from 1.9896 to 0.3456 failures per year. These results are also given in Table 30.

Table 29
Simple Radial System Relative
Ranking of Forced Hours of Downtime per Year

	$\lambda \cdot r$, Forced Hours of Downtime per Year
1. Electric utility	2.582
2. Transformer	1.0260*
3. 13.8 kV circuit breaker	0.2992*
4. 13.8 kV cable and terminations	0.1908*
5. Switchgear bus—insulated	0.0911
6. Switchgear bus—bare	0.0576
7. Disconnect switch	0.0220
8. 480 V circuit breaker	0.0108
9. 480 V circuit breaker (main)	0.0108
10. 480 V cable and terminations	0.0052
11. 480 V circuit breakers (5) (failed while opening)	0.0048
12. Protective relays (3)	0.0030
Total	4.3033

*Data for hours of downtime per failure are based upon *repair failed unit.*

Conclusion—Primary-Selective to the 13.8 kV Utility Supply. The use of primary-selective to the 13.8 kV utility supply with 9-minute manual switchover time reduces the forced hours downtime per year at the 480 volt point of use by about 50 percent, but the

Table 30
Simple Radial System and Primary-Selective System to 13.8 kV Utility Supply Reliability and Availability Comparison of Power at 480 V Point of Use

Distribution System	λ, Failures per Year	$\lambda \cdot r$, Forced Hours Downtime per Year
Example 1 Simple radial system	1.9896	4.3033
Example 2a Primary-selective system to 13.8 kV utility supply (with 9 min switchover after a supply failure)	1.9896	2.1291
Example 2b Primary-selective system to 13.8 kV utility supply (with switchover in less than 5 s after a supply failure)*	0.3456	1.8835

*Loss of 480 V power for less than 5 s is not counted as a failure.

failure rate is the same as for a simple radial system.

The use of automatic throwover equipment that could sense a failure of one 13.8 kV utility supply and switchover to the second supply in less than 5 seconds would give a 6:1 improvement in the failure rate at the 480 volt point of use (a loss of 480 volt power for less than 5 seconds is not counted as a failure).

7.1.8 *Example 3—Primary-Selective to Load Side of 13.8 kV Circuit Breaker.*

Description of Primary-Selective To Load Side of 13.8 kV Circuit Breaker. Figure 14 shows a one-line diagram of the power distribution system for primary-selective to load side of 13.8 kV circuit breaker. What are the failure rate and the forced hours downtime per year at the 480 volt point of use? *Example 3a*—Assume 9-minute manual switchover time. *Example 3b*—Assume automatic switchover can be accomplished in less than 5 seconds after a failure (loss of 480 volt power for less than 5 seconds is not counted as a failure).

Results—Primary-Selective to Load Side of 13.8 kV Circuit Breaker. The results from the reliability and availability calculations are given in Table 31.

Conclusions—Primary-Selective to Load Side of 13.8 kV Circuit Breaker. The forced hours downtime per year at the 480 volt point of use in Example 3 (Primary-Selective to Load Side of 13.8 kV Circuit Breaker) is about 10 percent lower than in Example 2 (Primary-Selective to 13.8 kV Utility Supply). The failure rate is about the same.

7.1.9 *Example 4—Primary-Selective to Primary of Transformer.*

Description of Primary-Selective to Primary of Transformer. Fig 15 shows a one-line diagram of the power distribution system for primary-selective to primary of transformer. What are the failure rate and the forced hours downtime per year at the 480 volt point of use? Assume 1 hour switchover time.

Results—Primary-Selective to Primary of Transformer. The results from the reliability and availability calculations are given in Table 32.

Conclusions—Primary-Selective to Primary of Transformer. The forced hours downtime per year at the 480 volt point of use in Example 4 (Primary-Selective to Primary of Transformer) is about 32 percent lower than for the Simple Radial System in Example 1. The failure rate is the same in Examples 4 and 1.

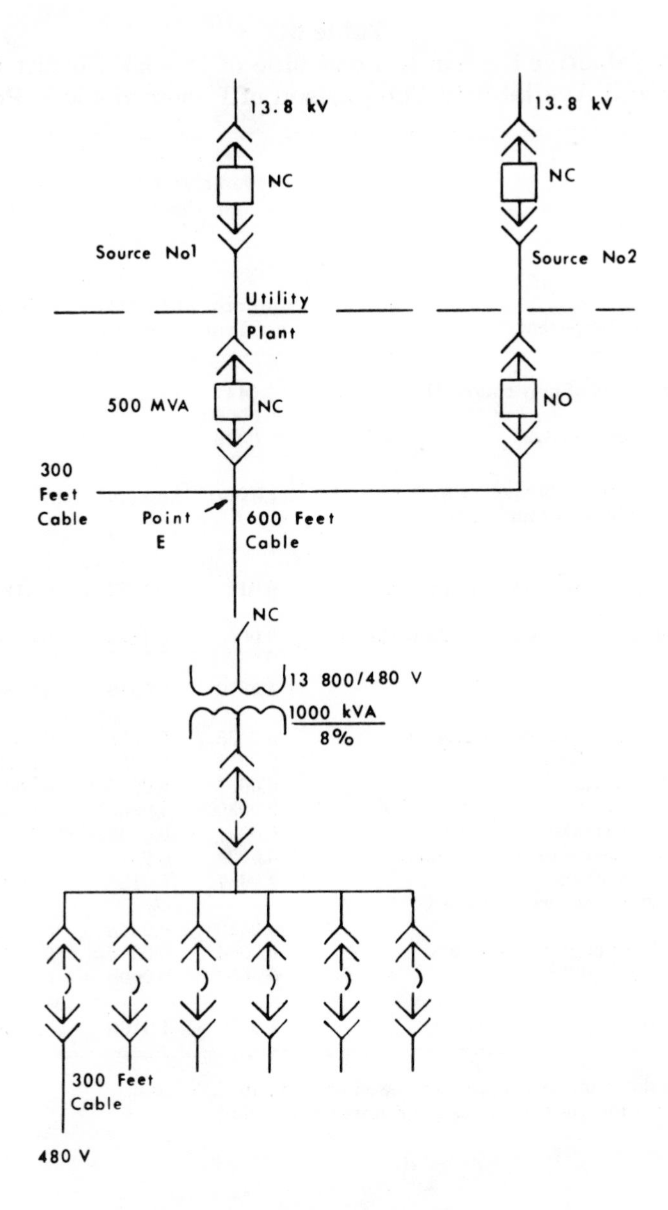

Fig 14
Primary Selective to Load
Side of 13.8 kV Circuit Breaker
Example 3

Table 31
Primary-Selective System to Load Side of 13.8 kV Circuit Breaker Reliability and Availability Comparison of Power at 480 V Point of Use

Component	Example 3a (9 min switchover time)		Example 3b (switchover in less than 5 seconds**)	
	λ, Failures per Year	$\lambda \cdot r$, Forced Hours of Downtime per Year	λ, Failures per Year	$\lambda \cdot r$, Forced Hours of Downtime per Year
13.8 kV power source (loss of only source 1)	1.644			
Protective relays (3)	0.0006			
13.8 kV metalclad circuit breaker	0.0036			
Total through 13.8 kV circuit breaker with 9 min switchover after a failure of source 1 (and source 2 is OK)	1.6482	0.2472		
Loss of both 13.8 kV power sources simultaneously	0.312	0.1622	0.312	0.1622
Switchgear bus—insulated (connected to 2 breakers)	0.0068	0.1822	0.0068	0.1822
Total to point E	1.9670	0.5916	0.3188	0.3444
Cable (13.8 kV); 900 ft, conduit below ground	0.0055	0.1458*	0.0055	0.1458*
Cable terminations (6) at 13.8 kV	0.0018	0.0450	0.0018	0.0450
Disconnect switch (enclosed)	0.0061	0.0220	0.0061	0.0220
Transformer	0.0030	1.0260*	0.0030	1.0260*
480 V metalclad circuit breaker	0.0027	0.0108	0.0027	0.0108
Switchgear bus—bare (connected to 7 breakers)	0.0024	0.0576	0.0024	0.0576
480 V metalclad circuit breaker	0.0027	0.0108	0.0027	0.0108
480 V metalclad circuit breakers (5) (failed while opening)	0.0012	0.0048	0.0012	0.0048
Cable (480 V); 300 ft, conduit above ground	0.0004	0.0044	0.0004	0.0044
Cable terminations (2) at 480 V	0.0002	0.0008	0.0002	0.0008
Total at 480 V output	1.9930	1.9196	.3448	1.6724

*Data for hours of downtime per failure are based upon *repair failed unit.*
**Loss of 480 V power for less than 5 s is not counted as a failure.

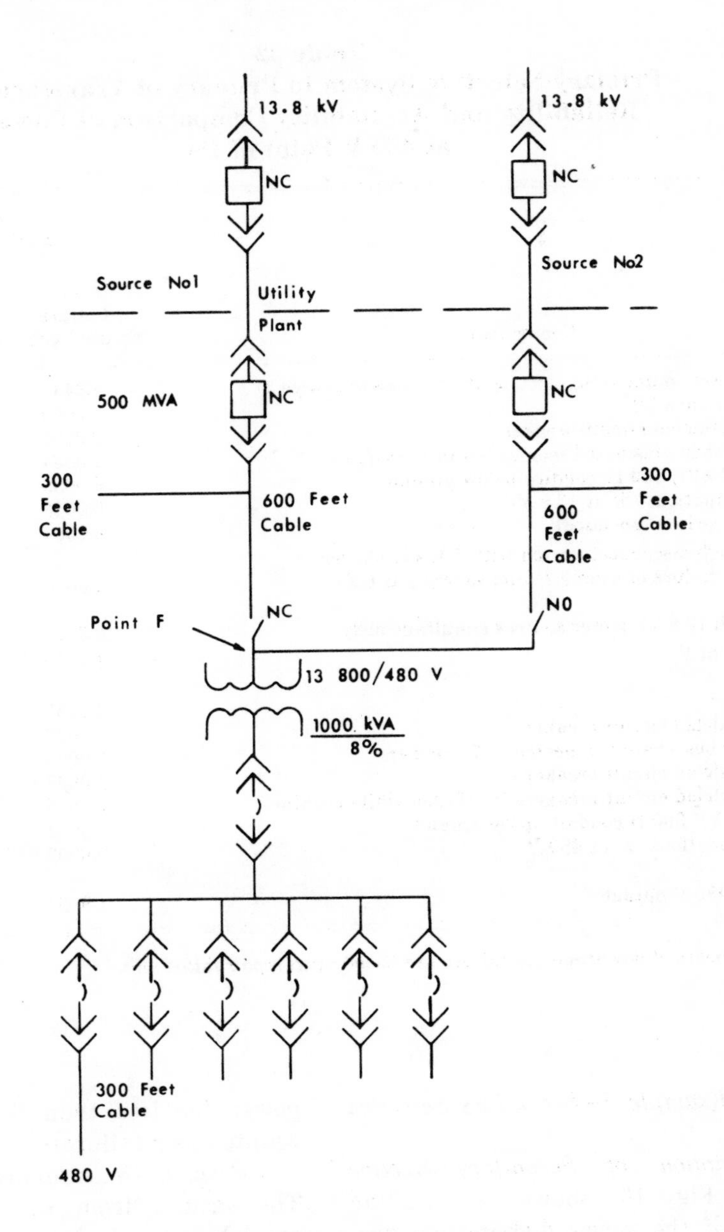

Fig 15
Primary Selective to Primary of Transformer Example 4

Table 32
Primary-Selective System to Primary of Transformer Reliability and Availability Comparison of Power at 480 V Point of Use

Component	Example 4 (switchover time 1 h)	
	λ, Failures per Year	λ·r, Forced Hours of Downtime per Year
13.8 kV power source from electric utility (loss of source 1)	1.644	
Protective relays (3)	0.0006	
13.8 kV metalclad circuit breaker	0.0036	
Switchgear bus—insulated (connected to 1 breaker)	0.0034	
Cable (13.8 kV); 900 ft, conduit below ground	0.0055	
Cable terminations (6) at 13.8 kV	0.0018	
Disconnect switch (enclosed)	0.0061	
Total through disconnect switch with 1 h switchover after a failure of source 1 (and source 2 is OK)	1.6650	1.6650
Loss of both 13.8 kV power sources simultaneously	0.312	0.1622
Total to point F	1.9770	1.8272
Transformer	0.0030	1.0260*
480 V metalclad circuit breaker	0.0027	0.0108
Switchgear bus—bare (connected to 7 breakers)	0.0024	0.0576
480 V metalclad circuit breaker	0.0027	0.0108
480 V metalclad circuit breakers (5) (failed while opening)	0.0012	0.0048
Cable (480 V); 300 ft conduit above ground	0.0004	0.0044
Cable terminations (2) at 480 V	0.0002	0.0008
Total at 480 V output	1.9896	2.9424

* Data for hours of downtime per failure are based upon *repair failed unit.*

7.1.10 *Example 5—Secondary Selective System.*

Description of Secondary-Selective System. Fig 16 shows a one-line diagram of the power distribution system for a secondary-selective system. What are the failure rate and forced hours of downtime per year at the 480 volt point of use? *Example 5a*—Assume 9-minute manual switchover time. *Example 5b*—Assume automatic switchover can be accomplished in less than 5 seconds after a failure (loss of 480 volt power for less than 5 seconds is not counted as a failure).

Results—Secondary-Selective System. The results from the reliability and availability calculations are given in Table 33.

Conclusions—Secondary-Selective System. The Simple Radial System in Example 1 had an average forced hours downtime per year that was 19 times larger than the Secondary-Selective System in Example 5b with automatic throwover in less than 5 seconds. The

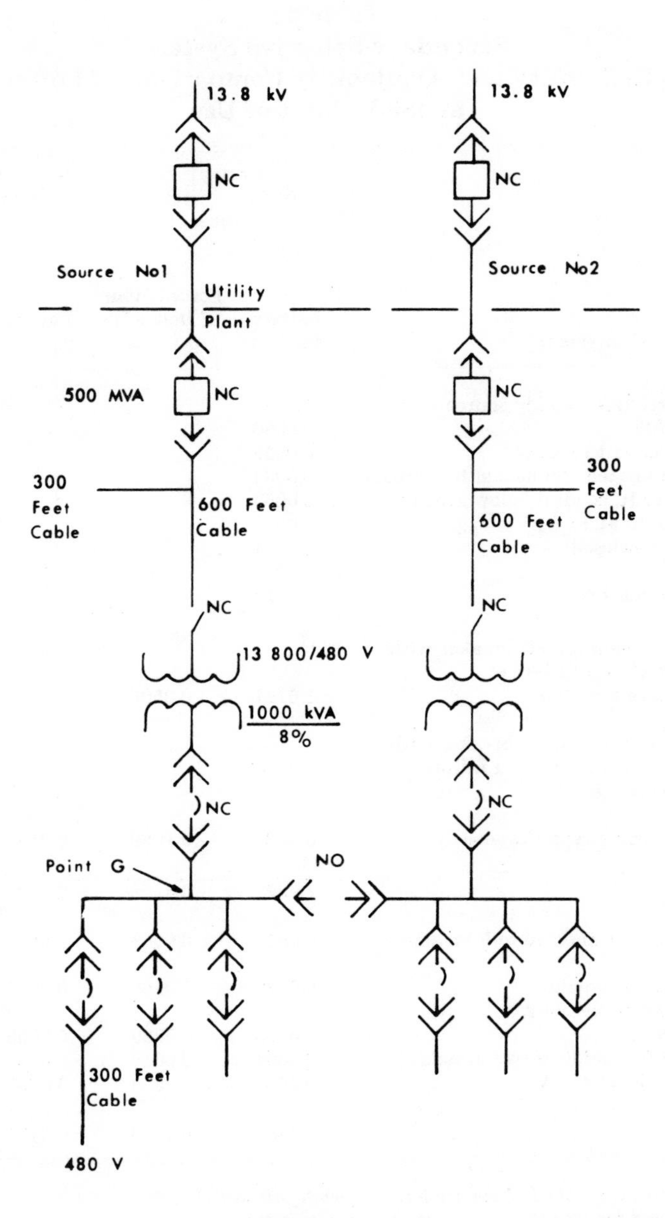

Fig 16
Secondary-Selective System
Example 5

Table 33
Secondary-Selective System
Reliability and Availability Comparison of Power at 480 V Point of Use

	Example 5a (9 min. switchover time)		Example 5b (switchover in less than 5 s**)	
Component	λ, Failures per Year	$\lambda \cdot r$, Forced Hours of Downtime per Year	λ, Failures per Year	$\lambda \cdot r$, Forced Hours of Downtime per Year
13.8 kV power source (loss of only source 1)	1.644			
Protective relays (3)	0.0006			
13.8 kV metalclad circuit breaker	0.0036			
Switchgear bus—insulated (connected to 1 breaker)	0.0034			
Cable (13.8 kV); 900 ft, conduit below ground	0.0055			
Cable terminations (6) at 13.8 kV	0.0018			
Disconnect Switch (enclosed)	0.0061			
Transformer	0.0030			
480 V metalclad circuit breaker	0.0027			
Total through 480 V main circuit breaker with 9 min switchover after a failure of source 1 (and source 2 is OK)	1.6707	0.2506		
Total through 480 V main circuit breaker with switchover in less than 5 s after a failure of source 1 (and source 2 is OK)			0	0
Loss of both power sources simultaneously	0.312	0.1622	0.312	0.1622
Total to point G	1.9827	0.4128	0.312	0.1622
Switchgear bus—bare (connected to 5 breakers)	0.0017	0.0408	0.0017	0.0408
480 V metalclad circuit breaker	0.0027	0.0108	0.0027	0.0108
480 V metalclad circuit breakers (2) (failed while opening)	0.0005	0.0020	0.0005	0.0020
Cable (480 V); 300 ft, conduit above ground	0.0004	0.0044	0.0004	0.0044
Cable terminations (2) at 480 V	0.0002	0.0008	0.0002	0.0008
Total at 480 V output	1.9882	0.4716	0.3175	0.2210

*Data for hours downtime per failure are based upon *repair failed unit.*
**Loss of 480 V power for less than 5 s is not counted as a failure.

failure rate of the Simple Radial System was six times larger than the Secondary-Selective System in Example 5b with automatic throwover in less than 5 seconds.

7.1.11 *Example 6—Simple Radial System with Spares.*

Description of Simple Radial System with Spares. Fig 12 shows a one-line diagram of the power distribution system for a simple radial system. What are the failure rate and forced hours of downtime per year of the 480 volt point of use if all of the following spare parts are available and can be installed as a replacement in these average times?

(1) 13.8 kV circuit breaker (inside plant only)—2.1 h

(2) 900 ft of cable (13.8 kV)—19 h

(3) 1000 kVA transformer—130 h

The above three "replace with spare" times were obtained from Table 24 and are the actual values obtained from the IEEE Reliability Survey of Industrial Plants [1]. The times are much lower than the "repair failed unit" times that were used in Examples 1 through 5.

Results—Simple Radial System with Spares. The results of the reliability and availability calculations are given in Table 34. They are compared with those of the Simple Radial System in Example 1 using average outage times based upon "repair failed unit."

Conclusions—Simple Radial System with Spares. The Simple Radial System with Spares in Example 6 had a forced hours downtime per year that was 22 percent lower than the Simple Radial System in Example 1.

7.1.12 *Overall Results from Six Examples.* The results for the six examples are compared in Table 35 which shows the failure rates and the forced hours downtime per year at the 480 volt point of use.

These data do not include outages for scheduled maintenance of the electrical equipment. It is assumed that scheduled maintenance will be performed at times when 480 volt power output is not needed. If this is not possible, then outages for scheduled maintenance would have to be added to the numbers shown in Table 35. This would affect a Simple Radial System much more than a Secondary-Selective System because of redundancy of electrical equipment in the latter.

7.1.13 *Discussion—Cost of Power Outages.* The forced hours of downtime per year is a measure of forced unavailability and is equal to the product of (failures per year) × (average hours) downtime per failure. The average downtime per failure could be called restorability and is a very important parameter when the forced hours of downtime per year are determined. The cost of power outages in an industrial plant is usually dependent upon both the failure rate and the restorability of the power system. In addition, the cost of power outages is also dependent on the "plant restart time" after power has been restored [5]. The "plant restart time" would have to be added to the "average downtime per failure" r, in Table 35 when cost versus reliability and availability studies are made in the design of the power distribution system.

The IEEE survey of industrial plants [1] found that the average "plant restart time" after a failure that caused complete plant shutdown was 17.4 hours. The median value was 4.0 hours.

7.1.14 *Discussion—Definition of Power Failure.* A failure of 480 volt power was defined in this guide as a complete loss of incoming power for more than 5

Table 34
Simple Radial System with Spares
Reliability and Availability Comparison of Power at 480 V Point of Use

Component	Example 1 Simple Radial			Example 6 Simple Radial With Spares		
	λ, Failures per Year	r, Hours of Downtime per Failure	$\lambda \cdot r$, Forced Hours of Downtime per Year	λ, Failures per Year	r, Hours of Downtime per Failure	$\lambda \cdot r$, Forced Hours of Downtime per Year
13.8 kV power source from electric utility	1.956		2.582	1.956		2.582
Protective relays (3)	0.0006		0.0030	0.0006		0.0030
13.8 kV metalclad circuit breaker	0.0036	83.1*	0.2292*	0.0036	2.1**	0.0076**
Switchgear bus—insulated (connected to 1 breaker)	0.0034		0.0911	0.0034		0.0911
Cable (13.8 kV); 900 ft, conduit below ground	0.0055	26.5*	0.1458*	0.0055	19.0**	0.1045**
Cable terminations (6) at 13.8 kV	0.0018		0.0450	0.0018		0.0450
Disconnect switch (enclosed)	0.0061		0.0020	0.0061		0.0220
Transformer	0.0030	342.*	1.0260*	0.0030	130.**	0.3900**
480 V metalclad circuit breaker	0.0027		0.0108	0.0027		0.0108
Switchgear bus—bare (connected to 7 breakers)	0.0024		0.0576	0.0024		0.0576
480 V metalclad circuit breaker	0.0027		0.0108	0.0027		0.0108
480 V metalclad circuit breakers (5) (failed while opening)	0.0012		0.0048	0.0012		0.0048
Cable (480 V); 300 ft, conduit above ground	0.0004		0.0044	0.0004		0.0044
Cable terminations (2) at 480 V	0.0002		0.0008	0.0002		0.0008
Total at 480 V output	1.9896		4.3033	1.9896		3.3344

*Data for hours of downtime per failure are based upon *repair failed unit.*
**Data for hours of downtime per failure are based upon *replace with spare.*

Table 35
Reliability and Availability Comparison at 480 V Point of Use for Several Power Distribution Systems

		Switchover in Less than 5 s***		Switchover Time 9 min			
Distribution System	Example	λ, Failures per Year	$\lambda \cdot r$, Forced Hours of Downtime per Year	λ, Failures per Year	$\lambda \cdot r$, Forced Hours of Downtime per Year	λ, Failures per Year	$\lambda \cdot r$, Forced Hours of Downtime per Year
Simple radial	1					1.9896	4.3033*
Simple radial with spares	6					1.9896	3.3344**
Primary-selective to 13.8 kV utility supply	2	0.3456	1.8835*	1.9896	2.1291*		
Primary-selective to load side of 13.8 kV circuit breaker	3	0.3448	1.6724*	1.9930	1.9196*		
Primary-selective to primary of transformer (1 h switchover)	4					1.9896	2.9424*
Secondary-selective	5	0.3175	0.2210*	1.9882	0.4716*		

*Data for hours downtime per failure are based upon *repair failed unit* for 13.8 kV circuit breaker, 13.8 kV cable, and transformer.
**Data for hours downtime per failure are based upon *replace with spare* for 13.8 kV circuit breaker, 13.8 kV cable, and transformer.
***Loss of 480 volt power for less than 5 s is not counted as a failure.

seconds. This is consistent with the results obtained from the IEEE survey of industrial plants [1], [1] which found a median value of 10 seconds for the "maximum length of power failure that will not stop plant production."

7.1.15 *Discussion—Electric Utility Power Supply.* Previous reliability studies [6][7][8] have drawn conclusions similar to those made in this section. All of these previous studies have identified the importance of two separate power supply sources from the electric utility. The Power System Reliability Subcommittee made a special effort to collect reliability data on double-circuit utility power supplies in the recent IEEE survey [2]. These data are summarized in Table 26 and were used in Examples 2 through 5. The two power sources in a double-circuit utility supply are not completely independent, and the reliability and availability analysis must take this into consideration. The importance of this point is shown in Table 36 where a reliability and availability comparison is made between the actual double-circuit utility power supply and the calculated value from two completely independent utility power sources. The actual double-circuit utility power supply has a failure rate more than 200 times larger than two completely independent utility power sources. The actual double-circuit utility power supply data came from the IEEE survey [2] and are based upon 77 outages in 246 unit-years of service at 45 plants with "all circuit breakers closed." This is a broad composite from many industrial plants in different parts of the country.

It is believed that utility supply failure rates vary widely in various locations. One significant factor in this difference is believed to be different exposures to lightning storms. Thus, average values for the utility supply failure rate may not be valid for any one location. Local values should be obtained, if possible, from the utility involved, and these values should be used in reliability and availability studies.

No examples are included here on the reliability and availability improvement that could be obtained by using local generation rather than purchased power from an electric utility. However, it is of interest to note the very high reliability of location generation equipment found in the IEEE "Reliability Survey of Industrial Plants."[1]

7.1.16 *Other Discussion.* The reliability and availability analysis in the six

[1]Numbers in brackets correspond to those in the References at the end of this Section.

Table 36
Comparison of Actual and Calculated Reliability and Availability of Double-Circuit Utility Power Supply (Failure Defined as Loss of Both Power Sources)

	λ, Failures per Year	$\lambda \cdot r$, Forced Hours of Downtime per Year
Actual single-circuit utility power supply from IEEE survey[2]	1.956*	2.582*
Actual double-circuit utility power supply from IEEE survey[2]	0.312*	0.1622*
Calculated-two utility power sources at 13.8 kV that are completely independent	0.0012**	0.0008**

*Taken from Table 26.
**Calculated using single-circuit utility power supply data and the formula for parallel reliability shown in Fig 11.

examples was done for 480 volt low-voltage power distribution systems. It is believed that 600-volt systems would have similar reliability and availability.

One of the assumptions made in the reliability and availability analysis is that the failure rate of the electrical equipment remains constant with age. It is believed that this assumption does not introduce significant errors in the conclusions. However, it is suspected that the failure rate of cables may change somewhat with age. In addition, data collected by the Edison Electric Institute on failures of power transformers above 2500 kVA show that the failure rate is higher during the first few years of service. The reliability data collected in the IEEE survey [1] did not attempt to determine how the failure rate varied with age for any electrical equipment studied.

A logical question asked very often is, how accurate are reliability and availability predictions? It is believed that the predicted failure rates and forced outage hours per year are at best only accurate to within a factor of 2 to what might be achieved in the field. However, the relative reliability and availability comparison of the alternative power distribution systems studied should be more accurate than 2:1.

The Rome Air Development Center of the United States Air Force has had considerable experience comparing the predicted reliability of Electronic Systems with the actual reliability results achieved in the field. These results [9] show that there is approximately a 12 percent chance that the field failure rate will be more than 2:1 worse than the reliability prediction made using a reliability handbook for electronic equipment such as [10]. It might be expected that the prediction of reliability of industrial power systems would have an accuracy similar to that obtained by the Air Force with electronic systems.

Some of the errors introduced when making reliability and availability predictions using published industry failure rates for the electrical equipment are:

(1) All details that could contribute to unreliability are not included in the study.

(2) Some of the contributions from human error may not be properly included.

(3) Equipment failure rates can be influenced by the adequacy of the preventive maintenance program [1],[11]. Contamination from the environment can also have an influence on equipment failure rates.

(4) Correct conclusions can be made from statistical analysis on the average. But some plants will never experience these average problems that are designed for. For example, several plants will never have a transformer failure.

In spite of these limitations, it is believed that reliability and availability analyses can be very useful in cost-reliability and cost-availability tradeoff studies during the design phase of the power distribution system.

7.1.17 *Spot Network.* A spot network would have a calculated reliability and availability approximately the same as the automatic throwover secondary-selective system[7],[8]. In addition it would have the benefit of no momentary outage in the event of a failure of any of the 13.8 kV cables or equipment since bus voltage is not lost on a spot network.

7.1.18 *Protective Devices other than Drawout Circuit Breakers.* The six examples in this chapter used drawout circuit breakers as protective devices. Other types of protective devices are also available for use on power systems. The examples in this chapter attempted to show how to make reliability and availability calculations. No attempt was made to study the effect on reliability and availability of different types of protective devices nor to draw conclusions that any particular type of protective device was more cost effective than another.

7.2 Cost Data Applied to Examples of Reliability and Availability Analysis of Common Low-Voltage Industrial Power Distribution Systems"

7.2.1 *Cost Evaluation of Reliability* and Availability Predictions" Cost evaluations are made of the reliability and availability predictions of four power distribution system examples from Chapter 7.1. The "Revenue Requirements Method" described in Section 2.2.3.1 is utilized in order to determine the most cost-effective system.

7.2.2 *Description of Cost Evaluation Problem.* Management insists that the engineer utilize an economic evaluation in any capital improvement program. The elements to be included and a method of mathematically equating the cost impact to be expected from electrical interruptions and downtimes against the cost of a new system were presented in Chapter 2.2. It was pointed out that there are several acceptable ways of accomplishing the detailed economic analysis for evaluation of systems with varying degrees of reliability. One of those considered acceptable, the Revenue Requirements (RR) Method was presented in detail, and this method will be used in the analysis of four examples.

The four example systems included are:

Example 1—Simple Radial System—Single 13.8 kV Utility Supply.

Example 2b—Primary-Selective to 13.8 kV Utility Supply (Dual)—Switchover Time Less than Five Seconds.

Example 4—Primary-Selective to Primary of Transformer—13.8 kV Utility Supply (Dual)—Manual Switchover in One Hour

Example 5b—Secondary-Selective with Switchover Time Less than Five Seconds.

Table 35 lists the expected failures per year and the average downtime per year for each of the examples. These data will be used to show which of the examples has the Minimum Revenue Requirement (MRR) making allowances for:

(1) Plant Startup Time

(2) Revenues Lost

(3) Variable Expenses Saved

(4) Variable Expenses Incurred

(5) Investment

(6) Fixed Investment Charges

One of the benefits of such a rigidly structured analysis is that the presentation is made in a sequential manner utilizing cost/failure data prepared with the assistance of management. With this arrangement the results of the evaluation are less likely to be questioned than if a less sophisticated method were used.

7.2.3 *Procedures for Cost Analyses.* Utilizing the single-line diagrams for the four examples a component quantity take-off of each system was made, and a present-day installed unit costs assigned for each component. In the case of dual 13.8 kV Utility Company's Supply, the basic cost of the second supply was estimated on the basis of a hypothetical case, assuming that a one-time cost only would be incurred. The extension of the costs results in the overall installed cost for each of the four examples. A summary of the installed costs is presented in Table 37. The total installed costs for each example are listed again after item (12) in Table 38.

The Revenue Requirements Method is used to calculate the total cost in dollars per year of both the "installed cost" and the "cost of unreliability" for the four examples. The methods for making these calculations are tabulated in Table 38. The reliability data and the assumed cost values used are described in the next two sections.

7.2.4 *Reliability Data for Examples.* Table 35 can be used to determine the failures per year, λ, and the "average hours downtime per failure," r, for each of the examples. The value of "r" is determined from dividing "$\lambda \cdot r$" by "λ." The values of "r" and "λ" for the four examples are shown after (1) and (10) respectively in Table 38.

7.2.5 *Assumed Cost Values.* The following common cost factors were assumed in 1976 for use in all four of the examples:

10 hours/failure—plant startup time after a failure, s

$8,000/hr—revenues lost per hour of plant downtime, g_p

$6,000/hr—variable expenses saved per hour of plant downtime, X_p

$20,000/failure—variable expenses incurred per failure, X_i

0.4 per year—fixed investment charge factor, F

These values are shown in Table 38 after (2), (4), (5), (8), and (13) respectively.

7.2.6 *Results and Conclusions.* The minimum revenue requirements for each of the four examples are shown in item (15) at the bottom of Table 38. Some of the conclusions that can be made are tabulated below:

Example 1—Simple Radial System. This system requires the least initial investment—$61,700—but its minimum revenue requirements of $112,877 per year is second highest of the four examples analyzed.

Example 2b—Primary-Selective to 13.8 kV Utility Supply (Dual) with Switchover Time Less than Five Seconds. This system requires an initial investment of $141,700 or 2.3 times that of the simple radial system; however, the minimum revenue requirement is $74,495 per year, which is the least of the four examples.

Based on the data presented, Example 2b would be selected since it has the least minimum revenue requirement.

Example 4—Primary-Selective to Primary of Transformer, 13.8 kV Utility Supply (Dual)–Manual Switchover Time of One Hour. This system shows next to highest initial cost of $171,900 and the highest minimum revenue requirement of $154,250 per year. A major contributor to the high MRR is the fact that while a dual system has been provided the utility supplies one hour manual switchover requirement increases the failure rate and downtime to account for its high MRR. If an automatic switchover were utilized, the example would be competitive with Example 2b.

Table 37
Installed Costs*

Item	Unit Cost	Example 1		Example 2b		Example 4		Example 5b	
		Simple Radial System Single 13.8 kV Utility Supply		Primary Selective** System to 13.8 kV		Primary-Selective System** to Primary of Transformer Utility Supply		Secondary-Selective** System	
		Quantity	Total Cost	Quantity	Total Cost	Quantity	Total Cost	Quantity	Total Cost
Utility Service Standby Charge				LS	$ 80 000	LS	$ 80 000	LS	$ 80 000
Basic Equipment									
High-voltage circuit breaker, each	$ 20 000	1	$ 20 000	1	20 000	2	40 000	2	40 000
High-voltage circuit cable linear feed	12	600	7 200	600	7 200	1 200	14 400	1 200	14 400
1000 kVA transformer with 2-position switch, each	20 000	1	20 000	1	20 000			2	40 000
1000 kVA transformer with 3-position switch, each	23 000 1					23 000			
1600A low-voltage circuit breaker, each	6 000	1	6 000	1	6 000	1	6 000	3	18 000
600A MCCB, each	2 500	1	2 500	1	2 500	1	2 500	1	2 500
Low-voltage cable, linear feet	20	300	6 000	300	6 000	300	6 000	300	6 000
Subtotal—basic equipment cost			61 700		61 700		91 900		120 900
Total cost			61 700		141 700		171 900		200 900

*All cost estimates were made in 1976.
**Estimates based on assumption that utility company alternate primary service will require 4 mi of 13.8 kV pole line and a 4000 kVA reserve capacity in the utility company's substation.

Table 38
Sample Reliability Economics Problem*

	Example 1	Example 2b	Example 4	Example 5b
	Simple Radial System Single 13.8 kV Utility Supply	Primary Selective to 13.8 kV Utility Supply****	Primary Selective**** Primary of Transformer	Secondary Selective System***
(1) r = component repair time or transfer time to restore service, whichever is less, hours per failure	2.16	5.45	1.48	0.69
(2) s** = plant startup time, hours per failure	10	10	10	10
(3) $r + s$ [items (1) + (2)]	12.16	15.45	11.48	10.69
(4) g_p** = revenues lost per hour of plant downtime, $/h	$8 000	$8 000	$8 000	$8 000
(5) x_p** = variable expenses saved, $/h	$6 000	$6 000	$6 000	$6 000
(6) $g_p - x_p$ [items (4) − (5)], value of lost production, $/hr	$2 000	$2 000	$2 000	$2 000
(7) $(g_p - x_p)(r + s)$ = [items (6) × (3)], $/failure	$24 320	$30 900	$22 960	$21 380
(8) x_i** = variable expenses incurred per failure, $/failure	$20 000	$20 000	$20 000	$20 000
(9) Items (7) + (8)	$44 320	$50 900	$42 960	$41 380
(10) λ = failure rate per year	1.99	0.35	1.99	0.32
(11) Items (9) × (10) = X, $/year	$88 197	$17 815	$85 490	$13 241
(12) C** = investment, $	$61 700	$141 700	$171 900	$200 900
(13) F** = fixed investment charge factor, per year	0.4	0.4	0.4	0.4
(14) CF = fixed investment charges, $/year	$24 680	$56 680	$68 760	$80 360
(15) G = $X+CF$, [items (11) + (14)], minimum revenue requirement, $/year	$112 877	$74 495	$154 250	$93 601
Economic choice		Example 2b		

*All cost estimates were made in 1976.
**Assumed values in this sample problem.
***Switchover time less than 5 s.
****Manual switchover time 1 h.

Example 5b—Secondary-Selective System with Switchover Time Less than Five Seconds. This system requires the highest initial investment—$200,900—and produces the next to the least minimum revenue requirements of $93,601 per year.

7.3 References [2]

[1] IEEE COMMITTEE REPORT. Report on Reliability Survey of Industrial Plants. *IEEE Transactions on Industry Applications*, Mar/Apr, Jul/Aug, Sept/Oct 1974, pp 213–252, 456–476, 681.

[2] IEEE COMMITTEE REPORT. Reliability of Electric Utility Supplies to Industrial Plants. *IEEE Technical Conference*, 75-CHO947-1-1A, pp 131–133.

[3] GARVER, D. P., MONTMEAT, F. E. and PATTON, A. D. Power System Reliability I—Measures of Reliability and Methods of Calculation. *IEEE Transactions on Power Apparatus and Systems*, July 1964, pp 727–737.

[4] PATTON, A. D. Fundamentals of Power System Reliability Evaluation, *IEEE Industrial & Commercial Power Systems Technical Conference*, Los Angeles, May 10–13, 1976.

[5] GANNON, P. E. Cost of Interruptions; Economic Evaluation of Reliability *IEEE Industrial and Commercial Power Systems Technical Conference*, Los Angeles, May 10–13, 1976.

[6] DICKINSON, W. H., GANNON, P. E., HEISING, C. R., PATTON, A. D., and McWILLIAMS, D. W. Fundamentals of Reliability Techniques as Applied to Industrial Power Systems, Conf Rec 1971 IEEE Ind. Comm. Power Syst Tech Conf 71C18-IGA, pp 10–31.

[7] HEISING, C. R. Reliability and Availability Comparison of Common Low-Voltage Industrial Power Distribution Systems. IEEE Trans Ind Gen Appl, Vol IGA-6, pp 416–424, Sept/Oct 1970.

[8] HEISING, C. R., and DUNKI-JACOBS, J. R. Application of Reliability Concepts to Industrial Power Systems. Conf Rec 1972 IEEE Industry Applications Society Seventh Annual Meeting, 72CH0-685-8-1A, pp 287–296.

[9] FEDUCCIA, A. J., and KLION, J. How Accurate Are Reliability Predictions. Rome Air Development Center, 1968 Annual Symposium on Reliability, IEEE Catalog Number 68C33-R, pp 280–287.

[10] Reliability Stress and Failure Rate Data for Electronic Equipment, MIL-HDBK-217A, Department of Defense, 1 December, 1965.

[11] WELLS, S. J. Electrical Preventive Maintenance. IEEE Industrial and Commercial Power Systems Technical Conference, Los Angeles, May 10–13, 1976.

[2]References [1] and [2] are reprinted in Appendixes A, B, and D.

8. Basic Concepts of Reliability Analysis by Probability Methods

8.1 Introduction

This chapter provides theoretical background for reliability analysis used in other chapters, Chapter 2 in particular. Some basic concepts of probability theory are discussed as these are essential to the understanding and development of quantitative reliability analysis methods. Definitions of terms commonly used in system reliability analysis are also included. The three methods discussed are, the cut set approach, the state space method and the network reduction technique.

Definitions. Some commonly used terms in system reliability analysis are defined here. These terms are used in the wider context of system reliability activities. Additional definitions more specifically related to power distribution systems are given in 2.1.3.

component. A piece of equipment, a line or circuit, or a section of a line or circuit, or a group of items which is viewed as an entity for purposes of reliability evaluation.

system. A group of components connected or associated in a fixed configuration to perform a specified function of distributing power.

failure. The termination of the ability of an item to perform a required function. (See 2.1.3 for a more detailed definition applicable to industrial and commercial power distribution systems.)

mean time between failure (MTBF). The mean exposure time between consecutive failures of a component. It can be estimated by dividing the exposure time by the number of failures in that period, provided that a sufficient number of failures have occurred in that period.

failure rate. The mean number of failures per unit exposure time of a component.

mean time to repair (MTTR). The mean time to repair a failed component. It can be estimated by dividing the summation of repair times by the number of repairs, and it is, therefore, practically the average repair time.

8.2 Basic Probability Theory

This section discusses some of the basic concepts of probability theory. An appreciation of these ideas is essential to the understanding and development of reliability analysis methods.

8.2.1 *Sample Space.* Sample space is the set of all possible outcomes of a phenomenon. For example, consider a system of three distribution links. Assuming that each link exists either in the operating or up state or failed or down state, the sample space is

S = (1U, 2U, 3U), (1D, 2U, 3U), (1U, 2D, 3U), (1U 2U, 3D), (1D, 2D, 3U), (1D, 2U, 3D), (1U, 2D, 3D), (1D, 2D, 3D)

Here *i*U, *i*D denote that the component *i* is up or down, respectively. The possible outcomes of a system are also called system states and the set of all possible system states is called system state space.

8.2.2 *Event.* Now in the example of 3 distribution links, the descriptions (1D, 2D, 3U), (1D, 2U, 3D), (1U, 2D, 3D), (1D, 2D, 3D) define an event that two or three lines are in the failed state. Assuming that a minimum of two lines are needed for successful system operation, this set of states also defines the system failure. The event A is, therefore, a set of system states and the event A is said to have occurred if the system is in a state that is a member of set A.

8.2.3 *Probability.* A simple and useful way of looking at the probability of an occurrence of the event in a large number of observations.

Consider, for example, that a system is energized at time $t=0$ and the state of the system is noted at time t. This is said to be one observation. Now if this process is repeated N times and the system is observed in the failed state N_f times, the probability of the system being in failed state at time t is,

$$P_f(t)=N_f/N \quad N \to \infty \qquad \text{(Eq 1)}$$

8.2.4 *Combinatorial Properties of Event Probabilities.* Certain combinatorial properties of event probabilities are useful in reliability analysis are discussed in this section.

a. *Addition rule of probabilities*

Two events A_1 and A_2 are mutually exclusive if they cannot occur together. For events A_1 and A_2 that are not mutually exclusive, that is, which can happen together,

$$P(A_1 \cup A_2)=P(A_1)+P(A_2) - P(A_1 \cap A_2) \qquad \text{(Eq 2)}$$

where

$P(A_1 \cup A_2)$ = Probability of A_1 or A_2, or both

$P(A_1 \cap A_2)$ = Probability of A_1 and A_2 happening together.

When A_1 and A_2 are mutually exclusive, they cannot happen together, that is, $P(A_1 \cap A_2)=0$, and therefore (Eq 2) reduces to:

$$P(A_1 \cup A_2)=P(A_1) + P(A_2) \qquad \text{(Eq 3)}$$

b. *Multiplication rule of probabilities*

If the probability of occurrence of event A_1 is affected by the occurrence of A_2, then A_1 and A_2 are not independent. The conditional probability of event A_1, given that event A_2 has already occurred, is denoted by $P(A_1 | A_2)$ and:

$$P(A_1 \cap A_2)=P(A_2)=P(A_1 | A_2) \qquad \text{(Eq 4)}$$

This is also used to calculate the conditional probability:

$$P(A_1 \mid A_2) = P(A_1 \cap A_2)/P(A_2) \quad \text{(Eq 5)}$$

When, however, events A_1 and A_2 are independent, that is, the occurrence of A_2 does not affect the occurrence of A_1:

$$P(A_1 \cap A_2) = P(A_1)\, P(A_2) \quad \text{(Eq 6)}$$

c. *Complementation.*
$\bar{A}_1$ is used to denote the complement of event A_1. The complement $\bar{A}_1$ is the set of states that are not members of A_1. For example, if A_1 denotes states indicating system failure, then the states not representing system failure make $\bar{A}_1$.

$$P(\bar{A}_1) = 1 - P(A_1) \quad \text{(Eq 7)}$$

8.2.5 *Random Variable.* A random variable can be defined as a quantity that assumes values in accordance with probabilistic laws. A discrete random variable assumes discrete values whereas a random variable assuming values from a continuous interval is termed continuous random variable. For example, the state of a system is a discrete random variable and the time between two successive failures is a continuous random variable.

8.2.6 *Probability Distribution Function.* Probability distribution function describes the variability of a random variable. For a discrete random variable X, assuming values x_i, the probability density function is defined by:

$$P_X(x) = P(X = x) \quad \text{(Eq 8)}$$

The probability density function for a discrete random variable is also called probability mass function and has the following properties:

(1) $P_X(x) = 0$ unless x is one of the values $x_0, x_1, x_2, \ldots$

(2) $0 \leqslant P_X(x_i) \leqslant 1$

(3) $\sum_i P_X(x_i) = 1$

Another useful function is the probability distribution function or cumulative distribution function. It is defined by:

$$F_X(x) = P(X \leqslant x) = \sum P_X(x_i), \quad x_i \leqslant x \quad \text{(Eq 9)}$$

The probability density function $f_X(x)$ [or simply $f(x)$] for a continuous random variable is so defined that:

$$P(a \leqslant X \leqslant b) = \int_a^b f(y)\, dy \quad \text{(Eq 10)}$$

If, for example, X denotes the time to failure, Eq 10 gives the probability that the failure will occur in the interval (a, b). The corresponding probability distribution function for a continuous random variable is:

$$F(x) = P(-\infty \leqslant X \leqslant x) = \int_{-\infty}^{x} f(y)dy \quad \text{(Eq 11)}$$

The function $f(x)$ has certain specific properties [1][1] including the following:

$$\int_{-\infty}^{\infty} f(x)\, dx = 1 \quad \text{(Eq 12)}$$

8.2.7 *Expectation.* The probabilistic behavior of a random varaiable is completely defined by the probability density function. It is often, however, desir-

[1]Numbers in brackets correspond to those in the References at the end of this Section.

able to have a single value characterizing the random variable. One such value is the expectation. It is defined by:

$$E(X) = \sum_i x_i P_X(x_i) \quad \text{for discrete random variable}$$

$$= \int_{-\infty}^{\infty} x\, f(x)\, dx \quad \text{for continuous random variable}$$

The expectation of X is also called the mean value of X and has a special relationship to the average value of X in that if random variable X is observed many times and the arithmetic average of X calculated, it will approach the mean value as the number of observations increases.

8.2.8 *Exponential Distribution.* There are several special probability distribution functions [1], but the one of particular interest in reliability analysis is the exponential distribution, having the probability density function:

$$f(x) = \lambda \exp(-\lambda x) \qquad \text{(Eq 13)}$$

where λ is a positive constant. The mean value of random variable X with exponential distribution is

$$d = \int_0^{\infty} x\lambda\, e^{-\lambda x}\, dx$$

$$= 1/\lambda \qquad \text{(Eq 14)}$$

Also the probability distribution

$$F(x) = \int_0^{x} \lambda e^{-\lambda y} dy$$

$$= 1 - e^{-\lambda x} \qquad \text{(Eq 15)}$$

If the time between failures obeys the exponential distribution, the mean time between failures is $d = 1/\lambda$, where λ denotes the failure rate of the component. It should be noted that the failure rate for exponential distribution and only exponential distribution is constant.

8.3 Reliability Measures. The term reliability is generally used to indicate the ability of a system to continue to perform its intended function. Several measures of reliability are described in the literature and some of the meaningful indexes for repairable systems, expecially power distribution systems, are described in this section.

1. *Unavailability.* Unavailability is the steady state probability that a component or system is out of service due to failures or scheduled outages. If only the failed state is considered, this term is called forced unavailability.

2. *Availability.* Availability is the steady state probability that a component or system is in service, satisfactorily performing its intended funtion. Numerically availability is the complement of unavailability, that is:

Availability = 1 − unavailability

3. *Frequency of system failure.* This index can be defined as the mean number of system failures per unit time.

4. *Expected failure duration.* This index can be defined as the expected or long-term average duration of a single failure event.

8.4 Reliability Evaluation Methods Numerical values for reliability measures can be obtained either by analytical methods or through digital simulation. Only the analytical techniques are discussed here and discussion on simulation approach can be found in [1]. The three methods described in this chapter are the state-space approach, network reduction and cut-set methods. The state space approach is very general but becomes cumbersome for relatively large systems. The network-reduction procedure is applicable when the system consists of series and parallel subsystems. The cut-set approach is becoming increasingly popular in the reliability analysis of transmission and distribution networks and has been primarily used in this book. The state space and network methods are discussed in this chapter for reference and for potential benefit to the users of this guide.

8.4.1 *Minimal Cut Set Approach.* The cut set approach can be applied to systems with simple as well as complex configurations and is a very suitable technique for the reliability analysis of power distribution systems. A cut set is a set of components whose failure alone will cause system failure and a minimal cut set has no proper subset of components whose failure alone will cause system failure. The components of a minimal cut set are in parallel since all of them must fail to cause system failure and various minimal cut sets are in series as any one minimal cut set can cause system failure.

A simple approach for the identification of minimal cut sets is described in Chapter 2 but more formal algorithms are also available in literature [1]. Once the minimal cut sets have been obtained, the reliability measures can be obtained by the application of suitable formulae [1], [2]. Assuming component independence and denoting the probability of failure of components in cut set C_i by $P(\bar{C}_i)$, the probability (unavailability) and the frequency of system failure for m minimal cut-sets are given by:

$$
\begin{aligned}
P_{\mathrm{f}} &= P(\bar{C}_1 \cup \bar{C}_2 \cup \bar{C}_3 \cup \cdots \cup \bar{C}_m) \\
&= P(\bar{C}_1) + P(\bar{C}_2) + \cdots \\
&\quad + P(\bar{C}_m) \qquad \binom{m}{1} \text{ terms} \\
&\quad - [P(\bar{C}_1 \cap \bar{C}_2) + \cdots \\
&\quad + [P(\bar{C}_i \cap \bar{C}_j)] i \neq j \quad \binom{m}{2} \text{ terms} \\
&\quad \vdots \\
&\quad (-1)^{m-1} P(\bar{C}_1 \cap \bar{C}_2 \cap \cdots \\
&\quad \cap \bar{C}_m) \qquad \binom{m}{m} \text{ terms}
\end{aligned}
\qquad \text{(Eq 16)}
$$

where $\bar{C}_1 \cap \bar{C}_2$, for example, denotes the failure of components of both the minimal cut-sets 1 and 2 and therefore $P(\bar{C}_1 \cap \bar{C}_2)$ means the probability of failure of all the components contained in C_1 and C_2, that is:

$$P(\bar{C}_1 \cap \bar{C}_2) = \prod P_{id}, \qquad i \in (C_1 \cup C_2)$$

where

P_{id} = probability of component i being in the failed state
$= r_i / (d_i + r_i)$
$= \lambda_i / (\lambda_i + \mu_i)$

d_i = MTBF of component i

λ_i = failure rate of component i
$= 1/d_i$

r_i = MTTR of component i

μ_i = repair rate of component i
$= 1/r_i$

$\prod$ = product

The frequency of failure is given by:

$$f_{\mathrm{f}} = P(\bar{C}_1)\, W_1 + P(\bar{C}_2)\, W_2 + \cdots + P(\bar{C}_m)\, W_m - [P(\bar{C}_1 \cap \bar{C}_2)\, W_{1,2} + P(\bar{C}_1 \cap \bar{C}_3)\, W_{1,3} + \cdots + P(\bar{C}_i \cap \bar{C}_j)\, W_{i,j}],\ i \neq j$$

$$\cdot$$
$$\cdot$$
$$\cdot$$

$$(-1)^{m-1} P(\bar{C}_1 \cap \bar{C}_2 \cap \cdots \cap \bar{C}_m)\, W_{1,2,\cdots,m} \quad \text{(Eq 17)}$$

where

$$W_{i,j} = \sum_{k \in C_i \cup C_j} \mu_k$$

The mean failure duration is given by:

$$d_{\mathrm{f}} = P_{\mathrm{f}} / f_{\mathrm{f}}$$

When the mean time between the failure of components is much larger than the mean time to repair or in other words, the component availabilities approach unity, Eqs 16 and 17 can be approximated [3] by simpler equations:

$$P_{\mathrm{f}} = \sum_{i=1}^{m} P(\bar{C}_i) = \sum_{i=1}^{m} P_{cs_i} \quad \text{(Eq 18)}$$

and

$$f_{\mathrm{f}} = \sum_{i=1}^{m} P(\bar{C}_i)\, W_i = \sum_{i=1}^{m} f_{cs_i} \quad \text{(Eq 19)}$$

where P_{cs_i} and f_{cs_i} are the probability and frequency of cut-set event *i*. Also,

$$d_{\mathrm{f}} = P_{\mathrm{f}} / f_{\mathrm{f}} = \sum_{i=1}^{m} P_{cs_i} \Big/ \sum_{i=1}^{m} f_{cs_i} = \sum_{i=1}^{m} f_{cs_i}\, r_{cs_i} \Big/ \sum_{i=1}^{m} f_{cs_i} \quad \text{(Eq 20)}$$

where

d_{f} = system mean failure duration

r_{cs_i} = mean duration of cut set event *i*

The application of Eqs 19 and 20 to power distribution systems is discussed in Chapter 2. The components in a minimal cut-set behave like a parallel system, and f_{cs_i} (assuming *n* components in C_i) can be computed as follows:

$$f_{cs_i} = \prod_{j=1}^{n} P_{j\mathrm{d}} \sum_{j=1}^{n} \mu_j \quad \text{(Eq 21)}$$

and

$$r_{cs_i} = 1 \Big/ \sum_{j=1}^{n} \mu_j \quad \text{(Eq 22)}$$

For example, for a cut-set having three components 1, 2, and 3:

$$f_{cs_i} = \frac{\lambda_1 \lambda_2 \lambda_3 (\mu_1 + \mu_2 + \mu_3)}{(\lambda_1 + \mu_1)(\lambda_2 + \mu_2)(\lambda_3 + \mu_3)} \simeq \lambda_1 \lambda_2 \lambda_3 (r_1 r_2 + r_2 r_3 + r_3 r_1),$$

assuming $\lambda_i << \mu_i$

and

$$r_{cs_i} = \frac{r_1 r_2 r_3}{r_1 r_2 + r_2 r_3 + r_3 r_1}$$

8.4.2 *State Space Approach.* The state-space method is a very general approach and can be used when the components are independent as well as for systems involving dependent failure and repair modes. The different steps of this approach are illustrated using a simple example of a component in series with two parallel components, as shown in Fig 17.

1. *Enumerate the possible system states.* Assuming each component can exist either in the operating state (U) or in the failed state (D) and that the components are independent, there are eight possible system states. These states are numbered 1 through 8 in Fig 18 and the description of the component states is indicated in each system state.

2. *Determine interstate transition rates.* The transition rate from s_i (that is, state i) to s_j is the mean rate of the system passing from s_i to s_j. For example, in Fig 18 the system can transit from s_1 to s_2 by the failure of component 1 and the repair of component 1 will put the system back into s_1. Therefore, the transition rate from s_1 to s_2 is λ_1 and the transition rate from s_2 is s_1 is μ_1.

3. *Determine state probabilities.* When the components can be assumed independent, state probabilities can be found by the product rule as indicated in Equation 6. When, however, statistical dependence is involved, a set of simultaneous equations needs to be solved to obtain state probabilities [1]. Only the independent case is discussed here and for this, say the probability of being in state 2 can be determined by:

$$P_2 = P_{1d}\, P_{2u}\, P_{3u} \qquad \text{(Eq 23)}$$

where

p_{iu} = probability of component i being up

$d_i/(d_i + r_i)$

$\mu_i/(\lambda_i + \mu_i)$

and

p_{id} = probability of component i being in failed state

$= r_i/(d_i + r_i)$

$= \lambda_i/(\lambda_i + \mu_i)$

4. *Determine reliability measures.* The states constituting the failure or success or any other event of interest are identified. For the system shown in Fig 17, if the links 2 and 3 are fully redundant, system failure can occur if either component 1 fails or component 2 and 3

Fig 17
One Component in a Series with Two Components in Parallel

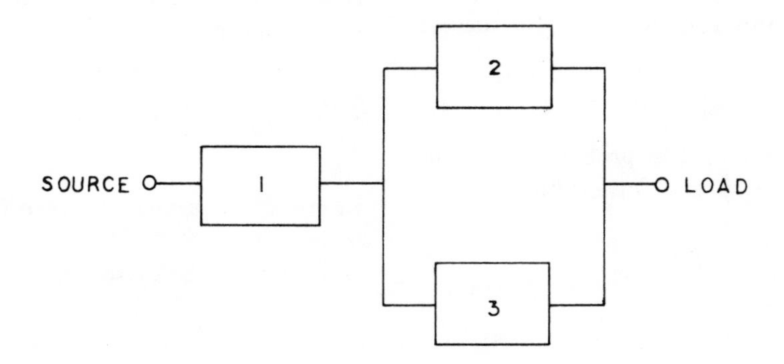

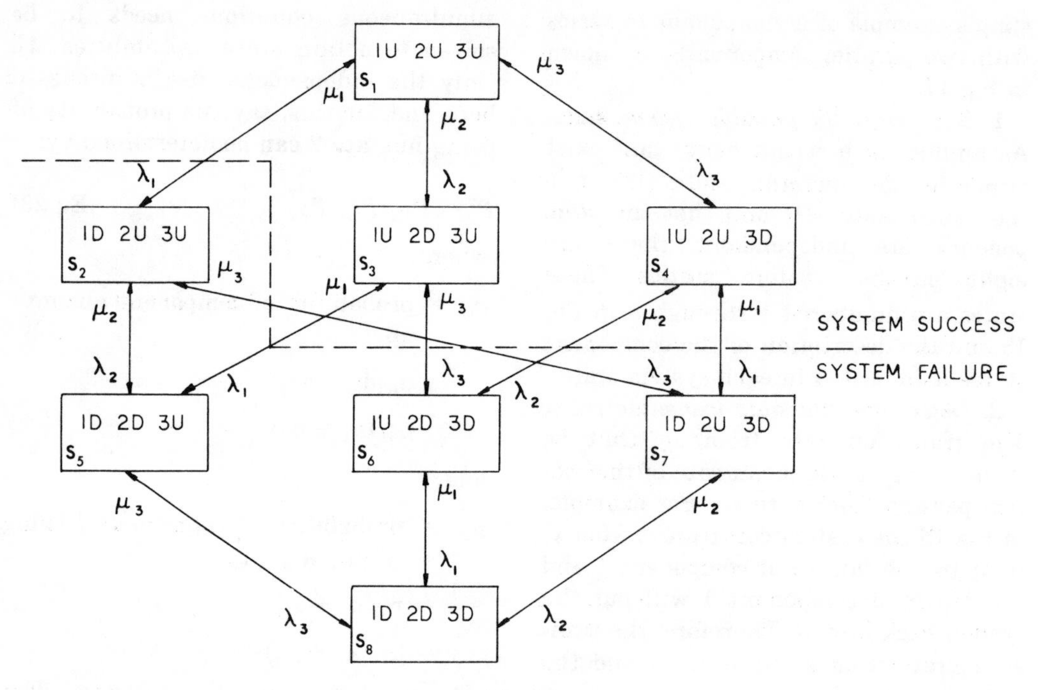

Fig 18
State Transition Diagram for the System Shown in Fig 17

fail or if all the components fail. The state space S as shown in Fig 18 is

$S = \{1, 2, 3, 4, 5, 6, 7, 8\}$

The subset A representing failure can be identified as:

$A = \{2, 5, 6, 7, 8\}$

and the subset representing the success states is

$S - A = \{1, 3, 4\}$

Unavailability or the probability of the system being in the failed state is now given by:

$$P_f = \sum_{i \in A} P_i \qquad \text{(Eq 24)}$$

where $i \in A$ indicates that summation is over all states contained in subset A.

Applying to our example,

$$P_f = P_2 + P_5 + P_6 + P_7 + P_8$$

where P_i can be found by the product rule as in Eq 23.

The frequency of system failure, that is, the frequency of encountering subset A can be computed by the following relationship:

$$f_f = \sum_{i \in (S-A)} P_i \sum_{j \in A} \lambda_{ij} \qquad \text{(Eq 25)}$$

where λ_{ij} = transition rate from state i to state j.

Applying Eq 25 to the system of Fig 17:

$$f_f = P_1\lambda_1 = P_3(\lambda_1 + \lambda_3) + P_4(\lambda_1 + \lambda_2)$$

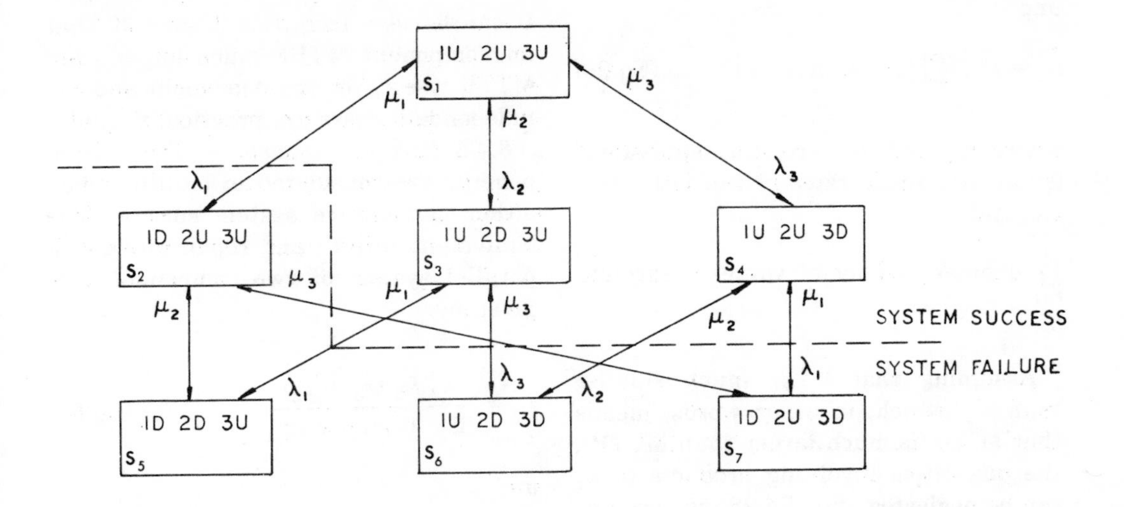

Fig 19
State Transition Diagram for the System Shown in Fig 17 when Components are not Independent

The mean failure duration can be obtained from P_f and f_f using:

$$d_f = P_f / f_f \quad \text{(Eq 26)}$$

In the preceding analysis, it has been assumed that the failure of a component does not alter the probability of failure of the remaining components. If, however, it is assumed that after the system failure, no further component failure will take place, the state transition diagram of Fig 18 will be modified as shown in Fig 19. Now once component 1 fails or components 2 and 3 fail, no further failure is possible. The probabilities in this case cannot be calculated by simple multiplication. It can be computed by solving a set of linear equations [1]. Once the state probabilities have been calculated, the remaining procedure is the same.

8.4.3 *Network Reduction.* The network reduction procedure is useful for systems consisting of series and parallel subsystems. The method consists in successively reducing the series and parallel structures by equivalent components. The knowledge of series and parallel reduction formulas is essential for the application of this technique.

8.4.4 *Series system.* The components are said to be in series when the failure of any one component causes system failure. It should be noted that the components do not have to be physically in series, it is the effect of failure that is important. Two types of series systems are discussed.

a. *Independent components.* For the series system of independent components, the failure and repair rate of the equivalent component are given by:

$$\lambda_s = \sum_{i=1}^{n} \lambda_i \quad \text{(Eq 27)}$$

and

$$\mu_s = \lambda_s / (\prod_{i=1}^{n} (1 + \lambda_i/\mu_i) - 1) \qquad \text{(Eq 28)}$$

where λ_s and μ_s are the equivalent failure and repair rates of the series system and

$\prod_{i=1}^{n}$ denotes product of values 1 through n.

Assuming that λ_i is much smaller than μ_i (which, in other words, means that MTBF is much larger than MTTR), the quantities involving products of λ_i can be neglected and Eq 28 reduces to:

$$r_s = 1/\mu_s = \sum_{i=1}^{n} r_i \lambda_i / \lambda_s \qquad \text{(Eq 29)}$$

b. *Components involving dependence.* When it is assumed that after the system failure no more components will fail, the equivalent failure and repair parameters are:

$$\lambda_s = \sum_{i=1}^{n} \lambda_i$$

and

$$r_s = \sum_{i=1}^{n} r_i \lambda_i / \lambda_s \qquad \text{(Eq 30)}$$

It can be seen from Eqs 29 and 30 that for component MTBF much larger than MTTR, the r_s for the dependent and the independent cases are practically equal.

8.4.5 *Parallel system.* Two components are considered in parallel when either can ensure system success. The equivalent failure and repair rates of a parallel system of two components are given by:

$$\lambda_p = \frac{\lambda_1 \lambda_2 (r_1 + r_2)}{1 + \lambda_1 r_1 + \lambda_2 r_2} \qquad \text{(Eq 31)}$$

and

$$\mu_p = \mu_1 + \mu_2 \qquad \text{(Eq 32)}$$

If $\lambda_1 r_1$ and $\lambda_2 r_2$ are much smaller than 1, then Eq 31 can be written as:

$$\lambda_p = \lambda_1 \lambda_2 (r_1 + r_2) \qquad \text{(Eq 33)}$$

8.6 References

[1] SINGH, C. and BILLINTON, R. *System Reliability Modelling and Evaluation*, London, Hutchinson Educational, 1977.

[2] SHOOMAN, L.M. *Probabilistic Reliability: An Engineering Approach*, New York, McGraw-Hill, 1968.

[3] SINGH, C. On the Behaviour of Failure Frequency Bounds, *IEEE Trans Reliability*, vol R-26, Apr 1977, pp 63–66.

Appendix A

Report on Reliability Survey of Industrial Plants

Part 1
Reliability of Electrical Equipment

Part 2
Cost of Power Outages, Plant Restart Time, Critical Service Loss Duration Time, and Type of Loads Lost Versus Time of Power Outages

Part 3
Causes and Types of Failures of Electrical Equipment, The Methods of Repair, and the Urgency of Repair

By
Reliability Subcommittee
Industry & Commercial Power Systems Committee
IEEE Industry Applications Society

W. H. Dickinson, *Chairman*

P. E. Gannon
M. D. Harris
C. R. Heising
W. J. Pearce
D. W. McWilliams
R. W. Parisian
A. D. Patton

Industrial and Comercial Power Systems Technical Conference
Institute of Electrical and Electronics Engineers, Inc
Atlanta, Georgia
May 13-16, 1973

Published by
IEEE Transactions on Industry Applications
Mar/Apr 1974

Report on Reliability Survey of Industrial Plants, Part I: Reliability of Electrical Equipment

IEEE COMMITTEE REPORT

Abstract—**An IEEE sponsored survey of electrical equipment reliability in industrial plants was completed during 1972. The results are reported from this survey which included a total of 1982 equipment failures that were reported by 30 companies covering 68 plants in nine industries in the United States and Canada.**

INTRODUCTION

A KNOWLEDGE of the reliability of electrical equipment is an important consideration in the design of power distribution systems for industrial plants. It is possible to make quantitative reliability comparisons between alternative designs of new systems and then use this information in cost-reliability tradeoff studies to determine which type of power distribution systems to use [1]-[10]. The cost of power outages at the various plant locations can be factored into the decision as to which type of power distribution system to use. These decisions can then be based upon total owning cost over the useful life of the equipment rather than first cost.

In 1969 a Reliability Working Group was formed under the Industrial Plants Power Systems Subcommittee, Industrial and Commercial Power Systems Committee. In 1972 the activity was changed to a Reliability Subcommittee under the same Committee. One of the major activities of the Reliability Working Group and the Reliability Subcommittee has been to conduct a survey of equipment reliability in industrial plants. This survey was conducted during the latter half of 1971 and the early part of 1972 and attempted to update a similar survey [11] which had been conducted eleven years ago. The results from the present survey contain data on failure rate and average downtime per failure for 74 equipment categories. The Reliability Subcommittee also felt that additional information was needed in the present survey beyond what was collected twelve years ago. Some of the additional information is the following:

1) cost of power outages of industrial plants;
2) plant restart time;
3) critical service loss duration time;
4) type of loads lost versus time of power outages;
5) repair or replacement time data;
6) repair urgency information;
7) causes and types of failures;
8) maintenance data and policies.

Paper TOD-73-158, approved by the Industrial and Commercial Power Systems Committee of the IEEE Industry Applications Society for presentation at the 1973 Industrial and Commercial Power Systems Technical Conference, Atlanta, Ga., May 13-16. Manuscript released for publication November 5, 1973.

Members of the Reliability Subcommittee of the IEEE Industrial and Commercial Power Systems Committee are W. H. Dickinson, *Chairman*, P. E. Gannon, M. D. Harris, C. R. Heising, D. W. McWilliams, R. W. Parisian, A. D. Patton, and W. J. Pearce.

It is not practical to publish all the results contained in the survey in a single paper. They will be presented in six separate parts. The first three parts are published at this time

Part 1: Reliability of Electrical Equipment;
Part 2: Cost of Power Outages, Plant Restart Time, Critical Service Loss Duration Time, and Type of Loads Lost Versus Time of Power Outages [11];
Part 3: Causes and Types of Failures, Methods of Repair, and Urgency of Repair [12].

A major part of the data in these three papers are presented in summary form. It is expected that the additional three papers will be presented at a later date and will contain further in-depth information where questions have been raised to point out the need for such data.

SURVEY FORM

The survey form is shown in Appendix A. Three types of cards were used for reporting the information.

Card type 1 asks for data on plant identification and other general plant information.

Card type 2 asks for data on a specific equipment class, including the total number of installed units, on their failure experience, on maintenance practices, and on estimated repair times of failed equipment.

Card type 3 asks for data on each individual failure reported on a card type 2.

It was necessary to provide definitions for "failure" and "repair time."

A *failure* is defined as any trouble with a power system component that causes any of the following to occur:

1) partial or complete plant shutdown, or below-standard plant operation;
2) unacceptable performance of user's equipment;
3) operation of the electrical protective relaying or emergency operation of the plant electrical system;
4) de-energization of any electric circuit or equipment.

A failure on a public utility supply system may cause the user to have either 1) a power interruption or loss of service, or 2) a deviation from normal voltage or frequency of sufficient magnitude or duration to disrupt plant production. A failure on an in-plant component causes a forced outage of the component, and the component thereby is unable to perform its intended function until it is repaired or replaced.

Repair time of a failed component or duration of a failure is the clock hours from the time of the occurrence of the failure to the time when the component is restored to service, either by repair of the component or by substitution with a spare component. It is not the time required to restore service to a load by putting alternate circuits into operation. It includes time for diagnosing the trouble, locating the failed component, waiting for parts, repairing or replacing, testing, and restoring the component to service.

Response to Survey

A total of 30 companies responded to the survey questionnaire, reporting data on 68 plants from nine industries in the United States and Canada as shown in Table I. There was a total of 1982 equipment failures reported in the survey; this included more than 620 000 unit-years of experience. Many of the plants reported data covering more than one year of experience.

Most of the data were reported to the IEEE Reliability Subcommittee during late 1971 and early 1972. Unfortunately, a downturn in the business cycle during this period of time caused many companies to reduce their work force and because of this fewer were able to participate in the survey than had been originally hoped.

Survey Data Preparation

All of the returned survey questionnaire forms were reviewed. An attempt was made to clarify any discrepancies that were detected. Usable data were punched onto IBM cards for use in data processing.

Statistical Analysis of Equipment Failures

Two equipment parameters are of prime importance in making system reliability studies. These parameters are 1) failure rate and 2) average outage duration or repair time. The best estimate for the failure rate of a particular type of equipment is the number of failures actually observed, divided by the total exposure time in unit-years, that is,

$$\hat{\lambda} = \frac{f}{T} \tag{1}$$

where

- $\hat{\lambda}$ best estimate of failure rate in failures per unit-year
- λ true failure rate
- f number of failures observed
- T total exposure time in unit-years.

Statements regarding the accuracy of failure rate estimates can be made through the use of confidence limits [10], [14]-[17]. Failure rate confidence limits are upper and lower values of failure rate such that the following equations hold:

$$\Pr\,[\lambda_L \geqq \lambda] = \frac{1-\gamma}{2} \tag{2}$$

$$\Pr\,[\lambda \geqq \lambda_U] = \frac{1-\gamma}{2} \tag{3}$$

where

- λ_L lower confidence limit of failure rate
- λ_U upper confidence limit of failure rate
- γ confidence interval (or confidence level).

A typical value often chosen for the confidence interval is 0.90. Once values for λ_L and λ_U are found, one can say that λ, whose best estimate is $\hat{\lambda}$, lies between λ_L and λ_U with 100γ percent confidence. Clearly the narrower the interval between λ_L and λ_U, the greater one's confidence that $\hat{\lambda}$ is a good estimate of λ, the true failure rate. Expressions for λ_L and λ_U are given as follows [17]:

$$\lambda_L = \frac{\chi^2 (1-\gamma)/2, 2f}{2T} \tag{4}$$

$$\lambda_U = \frac{\chi^2 (1+\gamma)/2, 2f+2}{2T} \tag{5}$$

where $\chi^2 p, n$ is the p percentage point of a chi-squared distribution with n degrees of freedom. $\chi^2 p, n$ is tabled in statistical handbooks.

By substituting the value of T from (1) into (4) and (5) we get

$$\lambda_L = \frac{\chi^2 (1-\gamma)/2, 2f}{2f}(\hat{\lambda}) \tag{6}$$

$$\lambda_U = \frac{\chi^2 (1+\gamma)/2, 2f+2}{2f}(\hat{\lambda}). \tag{7}$$

The deviation of the lower confidence level from $\hat{\lambda}$ in percent of $\hat{\lambda}$ is

$$\%\mathrm{dev}_L = 100\left(1 - \frac{\lambda_L}{\hat{\lambda}}\right). \tag{8}$$

Similarly, the deviation of the upper confidence level from $\hat{\lambda}$ in percent of $\hat{\lambda}$ is

$$\%\mathrm{dev}_U = 100\left(\frac{\lambda_U}{\hat{\lambda}} - 1\right). \tag{9}$$

Equations (6)-(9) were used to develop Fig. 1. These curves avoid the need of looking up $\chi^2 p,n$. Here λ_L and λ_U are plotted in terms of percent deviation from λ as a function of the observed number of failures.

The best estimate for the average outage duration or repair time for a particular type of equipment is simply the average of the observed outage durations. Confidence limit expressions for average outage durations are also available if the distributional nature of outage durations is known [17]. However, such expressions are not given here primarily because the average outage durations given in this paper are intended as a rough guide only. Equipment outage durations are believed to be more a function of the nature of a power system's operator than an inherent function of the equipment itself. Hence, average outage durations for equipment used in reliability studies should be values believed most reasonable for the particular system being studied.

The data from the survey contained information on the failure and repair characteristics of 217 categories of equipment. However, the number of observed failures for many equipment categories was too small to allow adequately accurate estimates of failure rates to be made. The Reliability Subcommittee felt that a minimum of eight to ten observed failures was required for "good" accuracy when estimating equipment failure rates (see Fig. 1). Therefore, whenever possible and reasonable from an engineering point of view, equipment categories having less than ten observed failures were combined with other categories so as to bring the number of observed failures in the combined category up to a minimum of ten. In some cases an equipment category with a large number of

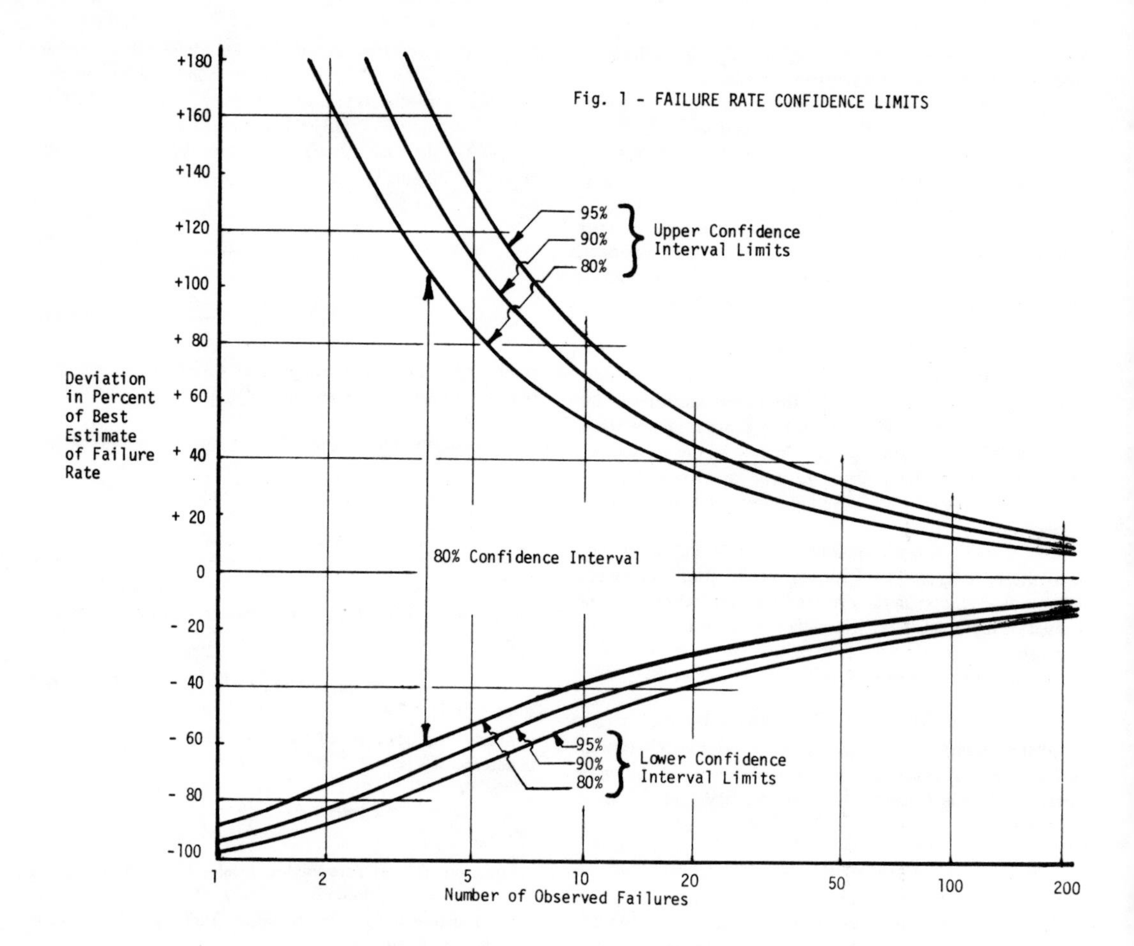

TABLE 1 - RESPONSE TO SURVEY QUESTIONNAIRE

Type of Industry	Number of Companies	Number of Plants
All Industry - USA & Canada	30*	68
Auto	0	0
Cement	0	0
Chemical	8	21
Metal	3	3
Mining	0	0
Petroleum	5	8
Pulp and Paper	1	1
Rubber & Plastics	3	3
Textile	1	3
Other Light Manufacturing	4	17
Other Heavy Manufacturing	1	2
Other	9	10
Foreign	1	1

*Some companies include more than one industry

observed failures was further subdivided. In most cases the equipment size attribute was eliminated by combining categories that were identical except for equipment size. These steps reduced the original 217 equipment categories to the 74 categories published in this paper. A total of 66 equipment categories have eight or more observed failures each; the other eight categories have between four and seven observed failures each.

SURVEY RESULTS OF EQUIPMENT FAILURES

Table 2 gives a summary of the "All Industry" equipment failure rate and equipment outage duration data for the 66 equipment categories that contain eight or more failures. The "actual hours downtime per failure" is based upon the actual outage data of the failed equipment; the "industry average" uses all equipment failures, and the "median plant average" uses all plants that reported actual outage time data on equipment failures.

The 1962 survey [11] contained equipment outage duration data on failures that have been challenged for two reasons.

1) Repairing a failed component may take much longer than replacing with a spare (for example, a large power transformer).

2) The urgency for repair is a significant factor in the outage time (low priority repairs may take days or weeks).

In order to help correct these deficiencies, two additional columns on "repair" and "replace with spare" were included in the survey and contain average estimated clock hours to fix failure during a 24-hour work day. These estimates are averaged over all the plants participating in the survey, even where there were no actual failures. These results are reported in Table 2 and are not included in the more detailed Tables 3-19.

Tables 3-19 give more detailed data on equipment failure rate and actual hours of equipment downtime per failure for 74 equipment categories; this includes the 66 equipment categories in Table 2 plus the eight equipment categories containing from four to seven failures. The additional detail includes

1) sample size in unit years;
2) number of failures;
3) number of plants reporting data;
4) additional data on actual hours of downtime per failure;
5) data for various industry groups where there were ten or more failures in that industry.

The data on average estimated clock hours to fix failure during 24-hour work day have been omitted from Tables 3-19.

The reliability data in Tables 14, 16, and 18 on cables, joints, and terminations represent a different look at the same data that are contained in Tables 13, 15, and 17. One set of tables looks at the type of insulation and the other set of tables looks at the application of the cable.

GENERAL COMMENTS AND DISCUSSION

A survey that collects data from many plants often contains errors. Some of the errors are due to a misinterpretation of the question by the respondent, and in other cases they can be caused by omission.

Many of the respondents apparently misinterpreted the question on "number of installed units" for double- or triple-circuit electric utility power supplies. In addition, there was some confusion on the outage time after a failure of a single circuit of a double- or triple-circuit utility power supply. See the separate discussion elsewhere in this paper on these points. These are the only known major problems of misinterpretation of survey questions.

It is suspected that the failure rate estimates may be biased on the high side due to the tendency of companies to report only on equipment that has actually experienced failures. In other words, some companies may have omitted submitting unit-years of experience data on equipment that had no failures. This factor may be partially balanced out by the belief that the companies that participated in the survey may be the ones that have the best maintenance programs and keep the best records and thus may have lower failure rates than the average.

It is expected that a future paper will contain a comparison of the equipment reliability from this survey with the results from the previous survey [11] that was published in 1962. A preliminary comparison has been made and shows the following overall conclusion for 1973 versus 1962.

1) The 1973 equipment failure rates are about 0.6 times the 1962 failure rates.

2) The 1973 average downtime per failure is about 1.6 times the 1962 average downtime per failure.

3) The product of failure rate times average downtime per failure is almost the same in 1973 as 1962.

Both of these parameters are within a factor of two; and this is often the best accuracy that can be expected from reliability data.

How accurate are the failure rates shown in Tables 2-19? Fig. 1 shows the upper and lower confidence limits of the failure rate versus the number of failures observed. It can be seen that ten failures has upper and lower confidence limits of +70 percent and -46 percent for a 90 percent confidence interval. It is possible to determine the upper and lower confidence limits for the failure rate data shown in Tables 3-19.

EXAMPLE OF CONFIDENCE LIMIT CALCULATION

The use of Fig. 1 to determine confidence limits will be illustrated with an example. Suppose that it is desired to compute confidence limits on the failure rate of liquid-filled transformers with voltage above 15 kV in the chemical industry. The desired confidence interval is 90 percent. From Table 4, $\hat{\lambda} = 0.0119$ failures per unit-year, and the number of observed failures is 19. Entering Fig. 1 with 19 observed failures and using the 90 percent confidence interval curves yields

$$\lambda_L = \hat{\lambda} - 0.34\hat{\lambda}$$
$$= 0.0119 - 0.0041 = 0.0078 \text{ failures per unit-year}$$
$$\lambda_U = \hat{\lambda} + 0.46\hat{\lambda}$$
$$= 0.0119 + 0.0055 = 0.0174 \text{ failures per unit-year.}$$

There is a 90 percent chance that the true failure rate lies between 0.0078 and 0.0174 failures per unit-year.

TABLE 2 - SUMMARY OF "ALL INDUSTRY" EQUIPMENT FAILURE RATE AND EQUIPMENT OUTAGE DURATION DATA FOR 66 EQUIPMENT CATEGORIES CONTAINING 8 OR MORE FAILURES

Equipment	Equipment Sub Class	Failure Rate-Failures per Unit-Year	Actual Hours Downtime per Failure: Industry Average	Actual Hours Downtime per Failure: Median Plant Average	Average Estimated Clock Hours to Fix Failure During 24 Hour Work Day: Repair Failed Component	Average Estimated Clock Hours to Fix Failure During 24 Hour Work Day: Replace with Spare
Electric Utility Power Supplies..	All..................................	0.643	1.33	1.04	-	-
" " " "	Single Circuit......................	0.537	5.66	5.10	-	-
" " " "	Double or Triple Circuit-All........	0.622	0.85	1.17	-	-
" " " "	Automatically Switched Over.......	0.735	0.59	0.93	-	-
" " " "	Manual Switchover.................	0.458	1.87	2.00	-	-
" " " " ...	Loss of All Circuits at One Time..	0.119	2.00	1.58	-	-
Transformers....................	Liquid Filled-All....................	0.0041	529.	219.	378.	73.4
"	601 - 15,000 Volts - All Sizes......	0.0030	174.	49.	382.	74.3
"	300-750 kVA........................	0.0037	61.0	10.7	49.0	3.7
"	751-2,499 kVA.....................	0.0025	217.	64.	297.	39.7
"	2,500 kVA & up....................	0.0032	216.	60.0	618.	150.
"	Above 15,000 Volts..................	0.0130	1076.	1260.	367.	71.5
"	Dry Type; 0 - 15,000 Volts...........	0.0036	153.	28.	67.	39.9
"	Rectifier; Above 600 Volts...........	0.0298	380.	80.	300.	20.0
Circuit Breakers...............	Fixed Type (incl. molded case) - All..	0.0052	5.8	4.0	31.7	4.5
" "	0 - 600 Volts - All Sizes..........	0.0044	4.7	4.0	6.0	2.0
" "	0 - 600 amps.....................	0.0035	2.2	1.0	4.0	2.0
" "	Above 600 amps...................	0.0096	9.6	8.0	8.0	2.0
" "	Above 600 Volts....................	0.0176	10.6	3.8	44.5	12.0
" "	Metalclad Drawout - All..............	0.0030	129.	7.6	54.2	3.9
" "	0 - 600 Volts - All sizes..........	0.0027	147.	4.0	47.2	2.9
" "	0 - 600 amps.....................	0.0023	3.2	1.0	75.6	1.2
" "	Above 600 amps...................	0.0030	232.	5.0	29.4	4.0
" "	Above 600 Volts....................	0.0036	109.	168.	62.4	5.2
Motor Starters..................	Contact Type; 0 - 600 Volts..........	0.0139	65.1	24.5	8.0	4.6
" "	Contact Type; 601 - 15,000 Volts......	0.0153	284.	16.0	23.6	13.8

TABLE 2 (Continued)

Equipment	Equipment Sub Class	Failure Rate - Failures per Unit-Year	Actual Hours Downtime per Failure: Industry Average	Actual Hours Downtime per Failure: Median Plant Average	Average Estimated Clock Hours to Fix Failure During 24 Hour Work Day: Repair Failed Component	Average Estimated Clock Hours to Fix Failure During 24 Hour Work Day: Replace with Spare
Motors	Induction; 0 - 600 Volts	0.0109	114.	18.3	50.2	13.0
"	Induction; 601 - 15,000 Volts	0.0404	76.0	91.5	71.4	19.7
"	Synchronous; 0 - 600 Volts	0.0007	35.3	35.3	32.0	10.0
"	Synchronous; 601 - 15,000 Volts	0.0318	175.	153.	146.	18.7
"	Direct Current - All	0.0556	37.5	16.2	69.0	5.3
Generators	Steam Turbine Driven	0.032	165.	66.5	234.	201.
"	Gas Turbine driven	0.638	23.1	92.0	190.	400.
Disconnect Switches	Enclosed	0.0061	3.6	2.8	50.1	13.7
Switchgear Bus - Indoor & Outdoor (Unit = Number of Connected Circuit breakers or Instrument Transformer Compartments)	Insulated; 601 - 15,000 Volts	0.00170	261.	26.8	41.0	66.0
	Bare; 0 - 600 Volts	0.00034	550.	24.0	41.5	24.5
	Bare; Above 600 Volts	0.00063	17.3	13.0	20.6	7.3
Bus duct - Indoor & Outdoor (Unit = One Circuit Foot)	All Voltages	0.000125	128.	9.5	12.9	6.0
Open Wire (Unit = 1,000 Circuit Feet)	0 - 15,000 Volts	0.0189	42.5	4.0	4.6	8.0
	Above 15,000 Volts	0.0075	17.5	12.0	8.0	-
Cable - All Types of Insulation. (Unit = 1,000 Circuit Feet)	Above Ground & Aerial					
	0 - 600 Volts	0.00141	457.	10.5	20.8	39.7
" "	601 - 15,000 volts - All	0.01410	40.4	6.9	26.8	60.4
" "	In Trays Above Ground	0.00923	8.9	8.0	49.4	119.
" "	In Conduit Above Ground	0.04918	140.	47.5	-	19.8
" "	Aerial Cable	0.01437	31.6	5.3	10.6	28.0
" "	Below Ground & Direct Burial					
" "	0 - 600 Volts	0.00388	15.0	24.0	-	26.8
" "	601 - 15,000 Volts - All	0.00617	95.5	35.0	20.4	26.8
" "	In Duct or Conduit Below Ground	0.00613	96.8	35.0	20.9	26.8
"	Above 15,000 Volts	0.00336	16.0	16.0	16.0	-

TABLE 2 (Continued)

Equipment	Equipment Sub Class	Failure Rate - Failures per Unit-Year	Actual Hours Downtime per Failure: Industry Average	Actual Hours Downtime per Failure: Median Plant Average	Average Estimated Clock Hours to Fix Failure During 24 Hour Work Day: Repair Failed Component	Average Estimated Clock Hours to Fix Failure During 24 Hour Work Day: Replace with Spare
Cable	601 - 15,000 Volts					
(Unit = 1,000 Circuit Feet)	Thermoplastic	0.00387	44.5	10.0	22.5	29.3
"	Thermosetting	0.00889	168.	26.0	27.2	55.2
"	Paper Insulated Lead Covered	0.00912	48.9	26.8	17.3	18.3
"	Other	0.01832	16.1	28.5	23.2	44.8
Cable Joints -All Types of Insul.	601 - 15,000 Volts					
" "	In Duct or Conduit Below Ground	0.000864	36.1	31.2	14.7	5.5
Cable Joints	601 - 15,000 Volts					
" "	Thermoplastic	0.000754	15.8	8.0	12.6	22.0
" "	Paper Insulated Lead Covered	0.001037	31.4	28.0	30.0	-
Cable Terminations - All Types of Insulation	Above Ground & Aerial					
" " " "	0 - 600 Volts	0.000127	3.8	4.0	8.0	8.0
" " " "	601 - 15,000 Volts - All	0.000879	198.	11.1	34.6	40.6
" " " "	Aerial Cable	0.001848	48.5	11.3	15.3	18.0
" " " "	in Trays Above Ground	0.000333	8.0	9.0	48.8	58.3
" "	In Duct or Conduit Below Ground 601 - 15,000 Volts	0.000303	25.0	23.4	28.8	30.0
Cable Terminations	601 - 15,000 Volts					
	Thermoplastic	0.004192	10.6	11.5	12.0	12.0
" "	Thermosetting	0.000307	451.	11.3	30.2	42.8
" "	Paper Insulated Lead Covered	0.000781	68.8	29.2	39.0	30.0
Miscellaneous	Inverters	1.254	107.	185.	5.0	8.0
"	Rectifiers	0.038	39.0	52.2	41.5	12.0

TABLE 3 - ELECTRIC UTILITY POWER SUPPLIES

Number of Plants in Sample Size	Sample Size Unit - Years	Number of Failures Reported	Industry	Equipment Sub Class	Failure Rate - Failures per Unit-Year	Actual Hours Downtime/Failure			
						Industry Average	Minimum Plant Average	Median Plant Average	Maximum Plant Average
30	314.4	202	All	All	0.643	1.33	*	1.04	24.0
7	70.8	38	"	Single Circuit	0.537	5.66	0.25	5.10	10.3
23	210.7	131	"	Double or Triple Circuit - All	0.622	0.85	*	1.17	24.0
17	140.2	103	"	Automativally Switched Over	0.735	0.59	*	0.93	6.00
6	54.6	25	"	Manual Switchover	0.458	1.87	1.82	2.00	24.0
23	210.7	25	"	Loss of All Circuits At One Time	0.119	2.00	*	1.58	6.00
7	64.8	20	Chemical	All	0.309	1.42	*	1.58	6.00
7	64.8	20	"	Double or Triple Circuit - All	0.309	1.42	*	1.58	6.00
6	60.1	20	"	Automatically Switched Over	0.333	1.42	*	1.58	6.00
3	46.5	10	Petroleum	All	0.215	6.80	0.33	4.95	9.57
2	18.5	49	Textile	All	2.649	0.28	0.014	2.17	4.33
2	18.5	49	"	Double or Triple Circuit - All	2.649	0.28	0.014	2.17	4.33
1	3.4	46	"	Automatically Switched Over	13.46	0.014	0.014	0.014	0.014
5	67.3	27	Other Light Manuf.	All	0.402	1.34	**	0.58	24.0
4	51.3	22	" " "	Double or Triple Circuit - All	0.429	1.51	**	0.79	24.0
3	27.3	15	" " "	Automatically Switched Over	0.549	0.51	**	0.04	1.46

* 19 cycles
** 2 seconds

TABLE 4 - TRANSFORMERS

Number of Plants in Sample Size	Sample Size Unit-Years	Number of Failures Reported	Industry		Failure Rate - Failures per Unit-Year	Actual Hours Downtime/Failure			
						Industry Average	Minimum Plant Average	Median Plant Average	Maximum Plant Average
33	15,210	63	All	Liquid Filled - All	0.0041	529.	2.0	219.	3744.
30	13,210	39	"	601-15,000 volts - All Sizes	0.0030	174.	2.0	49.	840.
12	3,002	11	"	300-750 kVA	0.0037	61.0	4.5	10.7	336.
18	6,040	15	"	751 - 2,499 kVA	0.0025	217.	2.0	64.0	840.
11	4,036	13	"	2,500 kVA & up	0.0032	216.	24.0	60.0	403.
12	1,848	24	"	Above 15,000 volts	0.0130	1076.	12.8	1260.	3744.
16	4,937	18	"	Dry Type; 0-15,000 volts	0.0036	153.	0.5	28.	720.
3	672	20	"	Rectifier, Above 600 volts	0.0298	380.	24.0	80.	867.
14	8,598	43	Chemical	Liquid Filled - All	0.0050	338.	8.0	168.	1800.
12	6,838	24	"	601-15,000 volts - All Sizes	0.0035	52.3	8.0	48.5	336.
7	3,274	10	"	300-750 kVA	0.0031	19.3	3.0	8.0	120.
9	1,601	19	"	Above 15,000 volts	0.0119	670.	12.8	708.	3600.
2	662	16	"	Rectifier; Above 600 volts	0.0242	425.	80.0	474.	867.
3	2,512	14	Petroleum	Liquid Filled - All	0.0056	843.	4.5	591.	1178.
3	2,334	10	"	601-15,000 volts - All Sizes	0.0043	244.	4.5	204.	403.

TABLE 5 - CIRCUIT BREAKERS

Number of Plants in Sample Size	Sample Size Unit-Years	Number of Failures Reported	Industry		Failure Rate - Failures per Unit-Year	Actual Hours Downtime/Failure Industry Average	Minimum Plant Average	Median Plant Average	Maximum Plant Average
16	9,501	49	All............	Fixed Type(includes molded case) - all	0.0052	5.8	0.5	4.0	72.0
12	8,990	40	"	0 - 600 volts - All Sizes..........	0.0044	4.7	0.5	4.0	11.0
9	7,643	27	"	0-600 amps......................	0.0035	2.2	0.5	1.0	9.0
4	1,347	13	"	Above 600 amps..................	0.0096	9.6	5.0	3.0	11.0
5	510	9	"	Above 600 volts....................	0.0176	10.6	1.5	3.8	72.0
28	40,770	124	"	Metalclad, Drawout - All............	0.0030	129.	0.3	7.6	890.
18	24,490	66	"	0-600 volts - All Sizes............	0.0027	147.	0.2	4.0	894.
11	11,270	26	"	0-600 amps......................	0.0023	3.2	0.2	1.0	4.0
13	13,220	40	"	Above 600 amps..................	0.0030	232.	0.2	5.0	894.
22	16,280	58	"	Above 600 volts....................	0.0036	109.	1.1	168.	883.
5	1,961	20	Chemical.......	Fixed Type(includes molded case) - All	0.0102	8.1	4.3	9.0	11.0
3	1,520	15	"	0-600 volts - All Sizes............	0.0099	9.5	5.0	9.0	11.0
2	937	13	"	Above 600 amps..................	0.0139	9.6	5.0	8.0	11.0
7	10,850	33	"	Metalclad, Drawout - All............	0.0030	83.7	5.8	97.7	576.
7	4,808	31	"	Above 600 volts....................	0.0064	89.3	6.3	97.7	576.
3	1,885	18	Petroleum.......	Fixed Type(includes molded case) - All	0.0095	5.8	1.0	4.0	72.0
2	1,817	17	"	0-600 volts - All Sizes............	0.0094	1.9	1.0	2.5	4.0
2	1,817	17	"	0-600 amps......................	0.0094	1.9	1.0	2.5	4.0
3	10,430	28	Textile.........	Metalclad, Drawout - All............	0.0027	289.	0.3	4.0	890.
3	9,655	25	"	0-600 volts - All Sizes............	0.0026	218.	0.3	4.0	894.
2	4,943	19	"	0-600 amps......................	0.0038	3.8	0.3	2.2	4.0

TABLE 6 - MOTOR STARTERS

Number of Plants in Sample Size	Sample Size Unit-Years	Number of Failures Reported	Industry	Equipment Sub Class	Failure Rate - Failures per Unit-Year	Actual Hours Downtime/Failure Industry Average	Minimum Plant Average	Median Plant Average	maximum Plant Average
			All..........	Contact Type					
9	4,522	63	"	0-600 volts......................	0.0139	65.1	1.0	24.5	75.5
15	6,518	100	"	601-15,000 volts..................	0.0153	284.	3.0	16.0	1440.
3	854	5	"	Circuit Breaker.....................	0.0059	2.8	2.8	2.8	2.8
7	5,340	14	Chemical......	Contact Type; 601-15,000 volts........	0.0026	298.	4.5	16.0	1323.
1	207	51	Metal.........	Contact Type; 0-600 volts............	0.2470	75.5	75.5	75.5	75.5
2	626	81	Petroleum.....	Contact Type; 601-15,000 volts........	0.1294	1440.	1440.	1440.	1440.

TABLE 7 - MOTORS

Number of Plants in Sample Size	Sample Size Unit-YEars	Number of Failures Reported	Industry	Equipment Sub Class	Failure Rate - Failures per Unit-Year	Actual Hours Downtime/Failure Industry Average	Minimum Plant Average	Median Plant Average	Maximum Plant Average
			All...............	Induction					
17	19,610	213	"	0-600 volts......................	0.0109	114.	0.5	18.3	312.
17	4,229	171	"	601-15,000 volts................	0.0404	76.0	3.3	91.5	191.
			"	Synchronous					
2	13,790	10	"...............	0-600 volts......................	0.0007	35.3	35.3	35.3	35.3
11	4,276	136	"	601-15,000 volts................	0.0318	175.	8.0	153.	360.
6	558	31	"	Direct Current....................	0.0556	37.5	4.0	16.2	139.
			Chemical..........	Induction					
6	9,638	50	"	0-600 volts............	0.0052	22.5	6.	10.3	45.7
8	2,819	122	"	601-15,000 volts........	0.0433	56.3	3.3	38.	191.
			"	Synchronous					
1	13,750	10	"	0-600 volts............	0.0007	35.3	35.3	35.3	35.3
4	1,201	52	"	601-15,000 volts........	0.0433	129.	25.8	113.	218.
			Petroleum.........	Induction					
3	6,467	146	"	0-600 volts............	0.0226	158.	120.	139.	159.
2	1,015	34	"	601-15,000 volts........	0.0335	139.	90.	119.	147.
			"	Synchronous					
2	2,826	78	"	601-15,000 volts........	0.0276	207.	167.	210.	254.
			Rubber & Plastics.	Induction					
3	161	12	" "	601-15,000 volts........	0.0748	144.	132	150.	168.
1	161	17	Textile..........	Direct Current...........	0.1056	9.4	9.4	9.4	9.4

TABLE 8 - GENERATORS

Number of Plants in Sample Size	Sample Size Unit-Years	Number of Failures Reported	Industry	Equipment Sub Class	Failure Rate - Failures per Unit-Year	Actual Hours Downtime/Failure: Industry Average	Minimum Plant Average	Median Plant Average	Maximum Plant Average
8	761.8	24	All............	Steam Turbine Driven......	0.032	165.	1.5	66.5	1080.
4	89.4	57	".............	Gas Turbine Driven........	0.638	23.1	5.0	92.0	720.
4	59.4	4	".............	Driven by Motor, Diesel, or Gas Engine.............	0.067	127.	121.	133.	144.
1	5.5	54	Petroleum.......	Gas Turbine Driven............	9.818	5.0	5.0	5.0	5.0

TABLE 9 - DISCONNECT SWITCHES

Number of Plants in Sample Size	Sample Size Unit-Years	Number of Failures Reported	Industry	Equipment Sub Class	Failure Rate - Failures per Unit-Year	Actual Hours Downtime/Failure: Industry Average	Minimum Plant Average	Median Plant Average	Maximum Plant Average
8	2,065	6	All..............	Open.....................	0.0029	183.	3.0	6.0	1080.
16	15,490	94	"	Enclosed................	0.0061	3.6	0.2	2.8	9.3
4	2,205	22	Chemical..........	Enclosed................	0.0100	6.0	2.0	5.1	6.5
1	4,293	61	Metal............	Enclosed................	0.0142	2.8	2.8	2.8	2.8

TABLE 10 - SWITCHGEAR BUS: INDOOR & OUTDOOR
(Unit = Number of Connected Circuit Breakers or Instrument Transformer Compartments)

Number of Plants in Sample Size	Sample Size Unit-Years	Number of Failures Reported	Industry	Equipment Sub Class	Failure Rate - Failures per Unit-Year	Actual Hours Downtime/Failure: Industry Average	Minimum Plant Average	Median Plant Average	Maximum Plant Average
12	11,740	20	All............	Insulated; 601-15,000 volts....	0.00170	261.	5.0	26.8	1613.
			"	Bare					
12	32,280	11	"	0-600 volts...............	0.00034	550.	2.0	24.0	2520.
5	20,560	13	"	Above 600 volts...........	0.00063	17.3	6.9	13.0	48.
5	4,003	15	Chemical........	Insulated; 601-15,000 volts.	0.00375	340.	18.0	26.8	1613.
			"	Bare					
3	17,270	10	"	Above 600 volts...........	0.00058	19.3	6.9	42.0	48.

TABLE 11 - BUS DUCT: INDOOR & OUTDOOR
(Unit = 1 Circuit Foot)

Number of Plants in Sample Size	Sample Size Unit-Years	Number of Failures Reported	Industry	Equipment Sub Class	Failure Rate - Failures per Unit-Year	Actual Hours Downtime/Failure			
						Industry Average	Minimum Plant Average	Median Plant Average	Maximum Plant Average
12	160,400	20	All..........	All Voltages..............	0.000125	128.	0.5	9.5	2160.

TABLE 12 - OPEN WIRE
(Unit = 1,000 Circuit Feet)

Number of Plants in Sample Size	Sample Size Unit-Years	Number of Failures Reported	Industry	Equipment Sub Class	Failure Rate-Failures per Unit-Year	Actual Hours Downtime/Failure			
						Industry Average	Minimum Plant Average	Median Plant Average	Maximum Plant Average
10	5,185	98	All..................	0-15,000 volts.........	0.0189	42.5	1.0	4.0	3600.
7	1,460	11	"	Above 15,000 volts........	0.0075	17.5	0.4	12.0	48.
3	292.6	10	Chemical............	0-15,000 volts...........	0.0342	606.	4.0	7.5	3600.
1	2,121	76	Petroleum...........	0-15,000 volts...........	0.0358	4.1	4.1	4.1	4.1

TABLE 13 - CABLE (ALL TYPES OF INSULATION)
(Unit = 1,000 Circuit Feet)

Number of Plants in Sample Size	Sample Size Unit-Years	Number of Failures Reported	Industry	Equipment SUb Class	Failure Rate-Failures per Unit-Year	Actual Hours Downtime/Failure Industry Average	Minimum Plant Average	Median Plant Average	Maximum Plant average
			All.................	Above Ground & Aerial					
10	5,692	8	"	0-600 volts......................	0.00141	457.	2.0	10.5	1802.
18	5,248	74	"	601-15,000 volts - All...........	0.01410	40.4	0.2	6.9	360.
7	1,517	14	"	In Trays Above Ground..........	0.00923	8.9	6.0	8.0	12.7
6	183	9	"	In Conduit Above Ground........	0.04918	140.	4.0	47.5	360.
11	3,548	51	"	Aerial Cable..................	0.01437	31.6	0.2	5.3	178.
			"	Below Ground & Direct Burial					
3	2,060	8	"	0-600 volts......................	0.00388	15.0	8.0	24.0	48.0
26	19,120	118	"	601-15,000 volts - All...........	0.00617	95.5	0.3	35.0	4320.
26	18,940	116	"	In Duct or Conduit Below Ground	0.00613	96.8	0.3	35.0	4320.
1	2,975	10	"	Above 15,000 volts..............	0.00336	16.0	16.0	16.0	16.0
			Chemical............	Above Ground & Aerial					
7	1,961	44	"	601-15,000 volts - All...........	0.02244	35.5	2.0	4.7	154.
3	1,137	11	"	In Trays Above Ground..........	0.00968	7.8	6.0	7.0	8.0
5	737	28	"	Aerial Cable..................	0.03800	47.1	2.0	4.7	178.
			"	Below Ground & Direct Burial					
10	11,420	70	"	601-15,000 volts - All...........	0.00613	53.0	2.6	25.0	514.
10	11,420	70	"	In Duct or Conduit Below Ground	0.00613	53.0	2.6	25.0	514.
			Petroleum..........	Above Ground & Aerial					
2	2,838	15	"	601-15,000 volts - All...........	0.00529	21.0	7.7	27.7	47.6
2	2,669	12	"	Aerial Cable..................	0.00450	23.1	7.7	53.8	100.
			"	Below Ground & Direct Burial					
2	981	23	"	601-15,000 volts - All...........	0.02345	94.0	26.8	69.7	113.
2	981	23	"	In Duct or Conduit Below Ground	0.02345	94.0	26.8	69.7	113.
1	2,975	10	"	Above 15,000 volts..............	0.00336	16.0	16.0	16.0	16.0

TABLE 14 - CABLE (ALL APPLICATIONS)
(Unit = 1,000 Circuit Feet)

Number of Plants in Sample Size	Sample Size Unit-Years	Number of Failures Reported	Industry	Equipment Sub Class	Failure Rate-Failures per Unit-Year	Actual Hours Downtime/Failure: Industry Average	Minimum Plant Average	Median Plant Average	Maximum Plant Average
			All..................	601-15,000 volts					
9	9,819	38	"	Thermoplastic.................	0.00387	44.5	2.0	10.0	178.
15	5,960	53	"	Thermosetting.................	0.00889	168.	0.2	26.0	4320.
10	7,126	65	"	Paper Insulated Lead Covered..	0.00912	48.9	0.3	26.8	120.
8	1,419	26	"	Other.........................	0.01832	16.1	0.7	28.5	168.
			Chemical.............	601-15,000 volts.					
7	9,158	36	"	Thermoplastic.................	0.00393	45.4	2.0	9.8	178.
3	2,578	26	"	thermosetting.................	0.01009	117.	17.3	202.	387.
4	937	26	"	Paper Insulated Lead Covered..	0.02774	10.7	2.6	25.0	120.
3	697	16	"	Other.........................	0.02297	18.3	8.0	9.0	168.
			Petroleum............	601-15,000 volts					
2	2,520	15	"	Thermosetting.................	0.00595	21.0	7.7	27.7	47.6
2	1,299	23	"	Paper Insulated Lead Covered..	0.01770	94.0	26.8	69.7	113.

TABLE 15 - CABLE JOINTS (ALL TYPES OF INSULATION)

Number of Plants in Sample Size	Sample Size Unit-Years	Number of Failures Reported	Industry	Equipment Sub Class	Failure Rate-Failures per Unit-Year	Actual Hours Downtime/Failure: Industry Average	minimum Plant Average	Median Plant Average	Maximum Plant Average
			All................	601-15,000 volts					
5	7,401	6	"	Above ground & Aerial..........	0.000811	20.3	8.0	16.5	48.0
12	40,500	35	"	In Duct or Conduit Below Ground	0.000864	36.1	1.0	31.2	160.
			Chemical...........	601-15,000 volts					
5	24,120	21	"	In Duct or Conduit Below Ground	0.000871	17.0	1.0	8.0	34.4

TABLE 16 - CABLE JOINTS (ALL APPLICATIONS)

Number of Plants in Sample Size	Sample Size Unit-Years	Number of Failures Reported	Industry	Equipment Sub Class	Failure Rate-Failures per Unit-Year	Actual Hours Downtime/Failure Industry Average	Minimum Plant Average	Median Plant Average	Maximum Plant Average
			All.................	601-15,000 volts					
5	27,860	21	"	Thermoplastic..................	0.000754	15.8	3.4	8.0	36.0
4	4,857	6	"	Thermosetting..................	0.001235	102.	14.0	60.0	160.
5	13,500	14	"	Paper Insulated Lead Covered...	0.001037	31.4	1.0	28.0	75.5
			Chemical............	601-15,000 volts					
4	22,900	20	"	Thermoplastic..................	0.000873	14.8	3.4	8.0	34.4

TABLE 17 - CABLE TERMINATIONS (ALL TYPES OF INSULATION)

Number of Plants in Sample Size	Sample Size Unit-Years	Number of Failures Reported	Industry	Equipment Sub Class	Failure Rate-Failures per Unit-Year	Actual Hours Downtime/Failure Industry Average	Minimum Plant Average	Median Plant Average	Maximum Plant Average
			All................	Above Ground & Aerial					
4	63,120	8	"	0-600 volts...................	0.000127	3.8	0.5	4.0	5.9
13	39,840	35	"	601-15,000 volts - All.......	0.000879	198.	1.0	11.1	728.
4	24,010	8	"	In Trays Above Ground......	0.000333	8.0	7.0	9.0	11.0
3	3,920	5	"	In Conduit Above Ground....	0.001276	1157.	24.0	732.	1440.
7	11,910	22	"	Aerial Cable..............	0.001848	48.5	1.0	11.3	84.4
			"	In Duct or Conduit Below Ground					
6	26,390	8	"	601-15,000 volts............	0.000303	25.0	16.0	23.4	34.5
			chemical..........	Above Ground & Aerial					
7	25,790	21	"	601-15,000 volts - All.......	0.000814	284.	7.0	11.2	728.
4	1,677	9	"	Aerial Cable..............	0.005367	14.6	9.0	13.7	24.0
			Petroleum..........	Above Ground & Aerial					
2	10,150	12	"	601-15,000 volts - All.......	0.001182	79.3	24.0	54.2	84.4
1	10,120	11	"	Aerial cable..............	0.001087	84.4	84.4	84.4	84.4

TABLE 18 - CABLE TERMINATIONS (ALL APPLICATIONS)

Number of Plants in Sample Size	Sample Size Unit-Years	Number of Failures Reported	Industry	Equipment Sub Class	Failure Rate-Failures per Unit-Year	Actual Hours Downtime/Failure Industry Average	Minimum Plant Average	Median Plant Average	Maximum Plant Average
			All............	601-15,000 volts					
2	2,385	10	"	Thermoplastic...............	0.004192	10.6	7.0	11.5	16.0
9	42,310	13	"	Thermosetting...............	0.000307	451.	9.3	11.3	1440.
5	20,490	16	"	Paper Insulated Lead Covered.	0.000781	68.8	16.0	29.2	82.6

TABLE 19 - MISCELLANEOUS

Number of Plants in Sample Size	Sample Size Unit-Years	Number of Failures Reported	Industry	Equipment Sub Class	Failure Rate-Failures per Unit-Year	Actual Hours Downtime/Failure Industry Average	Minimum Plant Average	Median Plant Average	Maximum Plant Average
5	3,164.	6	All...........	Fuses......................	0.0019	5.5	1.0	2.0	24.0
3	30,600.	6	"	Protective Relays...........	0.0002	5.0	0.5	3.8	7.2
3	11.2	14	"	Inverters..................	1.25	107.	2.1	185.	369.
3	314.	12	"	Rectifiers.................	0.0382	39.0	32.4	52.2	72.0
2	5.6	14	Chemical.......	Inverters..................	2.51	107.	2.1	185.	369.
1	16.8	10	Petroleum......	Rectifiers.................	0.5970	32.4	32.4	32.4	32.4

USER INSTRUCTIONS FOR IEEE SURVEY FORM ON
RELIABILITY OF ELECTRIC EQUIPMENT IN INDUSTRIAL PLANTS

(SPONSORED BY THE RELIABILITY WORKING GROUP,
INDUSTRIAL PLANTS POWER SYSTEMS SUBCOMMITTEE,
INDUSTRIAL AND COMMERCIAL POWER SYSTEMS COMMITTEE)

PURPOSE This survey is intended to collect data on failures that occur in in-plant electric equipment and in public utility electric power supplies that affect operations in industrial plants. We hope that these data will determine not only accurate failure rates and repair times on major classes of equipment, but will also give an insight into the causes of these failures in such a way that remedial recommendations may be formulated to reduce failures and to improve plant performance.

MAILING INSTRUCTIONS Mail all filled-out forms to the following address.

IEEE-IGA Reliability Working Group
Care of Assistant Professor A D Patton, Dept of Electrical Engineering
Texas A&M University
College Station, Texas 77843

DATA PROCESSING These forms will be given a confidential company code, and will then be key punched on cards for processing by a digital computer along with data collected from others. The computer will prepare a suitable report on failure rates, durations, and causes of failure.

ADDITIONAL INFORMATION The reverse side of the Survey Form asks for additional information. The following information should be filled in on the reverse side of the first page of data for each plant: company name, plant name, type and location, the name, address, and phone number of the individual submitting the data and/or the individual to whom questions about the data may be directed.

In addition, space is provided for remarks or clarifying comments on the data being reported. These comments should be filled in on all data sheets, if needed to clarify data.

DEFINITIONS

A component is a piece of equipment, a line or circuit, or a section of a line or circuit, or a group of items which is viewed as an entity.

A system is a group of components connected or associated in a fixed configuration to perform a specified function of generating, transmitting, or distributing power.

A failure is defined as any trouble with a power system component that causes any of the following to occur.

(1) Partial or complete plant shutdown, or below-standard plant operation
(2) Unacceptable performance of user's equipment
(3) Operation of the electrical protective relaying or emergency operation of the plant electrical system
(4) Deenergization of any electric circuit or equipment

A failure on a public utility supply system may cause the user to have either (1) a power interruption or loss of service, or (2) a deviation from normal voltage or frequency of sufficient magnitude or duration to disrupt plant production.

A failure on an in-plant component causes a forced outage of the component, and the component thereby is unable to perform its intended function until it is repaired or replaced.

Repair time of a failed component or duration of a failure is the clock hours from the time of the occurrence of the failure to the time when the component is restored to service, either by repair of the component or by substitution with a spare component. It is not the time required to restore service to a load by putting alternate circuits into operation.

It includes time for diagnosing the trouble, locating the failed component, waiting for parts, repairing or replacing, testing, and restoring the component to service.

Revision 3-4-71

WHD

2

USER INSTRUCTIONS FOR IEEE SURVEY FORM ON
RELIABILITY OF ELECTRIC EQUIPMENT IN INDUSTRIAL PLANTS
(SPONSORED BY THE RELIABILITY WORKING GROUP,
INDUSTRIAL PLANTS POWER SYSTEMS SUBCOMMITTEE,
INDUSTRIAL AND COMMERCIAL POWER SYSTEMS COMMITTEE)

GENERAL INSTRUCTIONS

THE SURVEY FORM The IEEE Survey Form 1`·1-70 is an input data form for a computer program. The data on these forms will be key punched onto computer cards and analyzed by the computer program.

CODED DATA The Survey Form asks for coded and uncoded data. It is necessary to refer to the instructions in filling in either. The following shows the columns on each card type that requires filling in a code.

CARD TYPE	COLUMNS REQUIRING CODES
1	1-10, 36
2	11-18, 33-36
3	25, 29, 30-53, 57, 58

It may happen that none of the codes shown fit the particular case being reported. For such cases, the "other" code should be used, by filling a "9" or a "99" in the space provided. "Other" means not otherwise classified. If this is done, explain on reverse side of page, referring to card type and column number.

EQUIPMENT CLASS A group of codes is used to specify an equipment class. An equipment class consists of a main code, two sub-class codes, a voltage code and a size code. These are explained in the instructions. For the example shown on the filled-out form, this code is as follows.

CLASS	CODE	DESCRIPTION
Main	20	= transformer
Sub 1	4	= power
Sub 2	34	= liquid filled
Voltage	2	= 601-15,000 volts primary
Size	3	= 300-750 kVA

The above coded equipment class covers all liquid-filled power transformers, with a primary voltage of 601-15,000 volts and rated 300-750 kVA. Any transformer in the plant that does not fit this example is a different classification and requires a different coding. Thus, a 5000 kVA power transformer, liquid filled, 13.8 kV primary voltage would be coded 20-4-34-2-5.

CARD-TYPES The Survey Form asks for three types of information under the headings CARD-TYPE 1, CARD-TYPE 2, and CARD-TYPE 3.

In general, CARD-TYPE 1 asks for data on plant identification and other general plant information.

CARD-TYPE 2 asks for data on a specific equipment class, including the total number of installed units, on their failure experience, on maintenance practices, and on estimated repair times of failed equipment. The total installed units and their failure experience is the most essential data asked for.

CARDS-TYPE 3 asks for data on each individual failure reported on a CARD-TYPE 2.

A typical plant might have as many as, say 30 different equipment classes. These 30 equipment classes might have, for example 10 different failures. To report this information requires 30 pages of the Survey Form, one for each different equipment class. CARD-TYPE 1 is filled in completely on the first page and partly thereafter. CARD-TYPE 2 is filled in on each page. CARDS-TYPE 3 are filled in 10 times, once for each failure, if any.

CARD-TYPE 1 CARD-TYPE 1 is used to identify the reporting company and plant of that company and to give general information about that plant. The first 10 columns on this card are to be repeated by the key puncher onto CARD-TYPE 2 and CARDS-TYPE 3 for identification purposes.

Only one CARD-TYPE 1 is used by the computer program. However, we ask that on each page of the IEEE Survey Form that the first 7 columns be filled-in in case the filled-out survey forms become separated.

Fill in Items 1-8 on reverse side of first page of data for each plant.

ALL CARD TYPES Fill in CARD-TYPE, column number, and remarks or comments on reverse side, if any, on all data cards.

USER INSTRUCTIONS FOR IEEE SURVEY FORM ON
RELIABILITY OF ELECTRIC EQUIPMENT IN INDUSTRIAL PLANTS
(SPONSORED BY THE RELIABILITY WORKING GROUP,
INDUSTRIAL PLANTS POWER SYSTEMS SUBCOMMITTEE,
INDUSTRIAL AND COMMERCIAL POWER SYSTEMS COMMITTEE

CARD-TYPE 2 The second or CARD-TYPE 2 is used to report on each different equipment class in the plant. A typical plant might have a one type of utility supply, and several different classes each of transformers, circuit breakers, cables, etc. These different classes are shown in Columns 11-18. These Columns 11-18 are to be repeated by the key puncher on all CARDS-TYPE 3. There will be as many CARDS-TYPE 2 as there are different equipment classes.

Each CARD-TYPE 2 is used to report (1) the total number installed of one equipment class and the total number of failures experienced (if any) of that equipment class.

In addition, each CARD-TYPE 2 is used to report on maintenance practices and estimated repair times. These are your best estimate of repair times. These estimated times will be used if actual repair times are not known, or if actual repair times are much different from the average for some special reason which is unlikely to recur. We prefer to use actual data if available.

These data are to be left blank for failures on the utility power supply, since this information is not normally available.

CARD-TYPE 3 The third or CARD-TYPE 3 is used to report on actual data for each failure reported on a corresponding CARD-TYPE 2. Thus, associated with each CARD-TYPE 2 is a set of CARDS-TYPE 3. The number of CARDS-TYPE 3 will be the same as the number of failures (column 31) reported on CARDS-TYPE 2, for example, if a CARD-TYPE 2 has a 3 in Column 31, then 3 CARDS-TYPE 3 should be filled in.

Each CARD-TYPE 3 reports specific information on one failure, such as failure duration, urgency of repair, cause of failure, loads affected by the failure, and effect of failure on plant operations.

RIGHT-ADJUSTMENT OF DATA In filling in data, numbers should be right-adjusted, that is, they must end in the right-hand column of the assigned field. This means that if, for example, the survey form provides 3 columns to insert data but a two-digit number is to be inserted in the space available, then the number should be filled into the two right-hand columns.

SAMPLE FILLED-OUT FORM Refer to the attached sample filled-out form. This gives an example of a report on one class of transformers with two failures.

7) DATE *3 - 4 - 71*

SAMPLE

IEEE SURVEY FORM 11-1-70

PAGES *15* PAGE *4*

RELIABILITY OF ELECTRIC EQUIPMENT IN INDUSTRIAL PLANTS

CARD - TYPE 1

(REFER TO SURVEY FORM INSTRUCTIONS)
(NOTE - * REFERS TO CODED DATA)

COM-PANY CODE	PLANT*					PLANT OPERATING SCHEDULE		ESTIMATED PLANT OUTAGE COST, $		PLANT MAX. DEMAND AT PLANT DESIGN CAPACITY, KW	PLANT RESTART TIME, HOURS	CRITICAL SERVICE LOSS DURATION		CARD TYPE	CARD NO.
	NO.	TYPE	LOCATION	CLIMATE	ATMOSPHERE	HR. PER DAY	DAYS PER WK.	PER FAILURE	PER HR. DOWNTIME			NO. OF UNITS	UNITS*		
1	4	6	8	9	10	11	13	15	20	25	31	33	36	79	80
GEL	1	10	1	5	1	8	5	4000	2000	54000	2	10	4	1	1

CARD - TYPE 2

EQUIPMENT CLASS*					PERIOD COVERED BY THIS REPORT				NO. OF INSTALLED UNITS	NUMBER OF FAILURES	AVERAGE AGE*	MAIN.* TENANCE		ESTIMATED CLOCK HOURS TO REPAIR A FAILURE				CARD TYPE	CARD NO.
					FROM		TO							REPAIR FAILED COMPONENT		REPLACE WITH SPARE			
MAIN	SUB 1	SUB 2	VOLTAGE	SIZE	MO.	YR.	MO.	YR.				NORMAL CYCLE, MO.	QUALITY	24-HR. PER DAY	8-HR. PER DAY	24-HR. PER DAY	8-HR. PER DAY		
11	13	15	17	18	19	21	23	25	27	31	33	34	36	37	41	45	48	79	80
20	4	34	2	3	1	66	10	70	120	2	3	3	2	100	300	14	48	2	1

CARDS - TYPE 3

FAILURE															LOADS LOST*					% PRODUCTION LOST*	PLANT OUTAGE DURATION		SERVICE RESTORED*	CARD TYPE	CARD NO.
NUMBER	DATE		FOREWARNING*	DURATION		REPAIR METHOD*	REPAIR URGENCY*	MO. SINCE LAST MAINTAINED*	DAMAGED PART*	TYPE*	RESPONSI-BILITY*	INITIATING CAUSE*	CONTRIBUTING CAUSE*	CHARACTER-ISTICS*	COMPUTER	MOTOR	LIGHTING	SOLENOID	OTHER		NO. OF UNITS	UNITS*			
	MO.	YR.		NO. OF UNITS	UNITS*																				
19	21	23	25	26	29	30	32	34	36	38	40	42	44	46	48	49	50	51	52	53	54	57	58	79	80
1	9	69		60	2	2	1	2	9	1	1	4	99	15	1	1	1	1	1	2	4	4	4	3	1
2	8	70		180	2	1	1	3	2	1	5	99	10	6	1	1	0	9	1	2	4	4	4	3	2
3																								3	3
4																								3	4
5																								3	5
6																								3	6
7																								3	7
8																								3	8
9																								3	9
10																								3	0

USER INSTRUCTIONS FOR CARD-TYPE 1

CARD - TYPE 1

(REFER TO SURVEY FORM INSTRUCTIONS)
(NOTE - * REFERS TO CODED DATA)

COM-PANY CODE	PLANT*					PLANT OPERATING SCHEDULE		ESTIMATED PLANT OUTAGE COST, $		PLANT MAX. DEMAND AT PLANT DESIGN CAPACITY, KW	PLANT RESTART TIME, HOURS	CRITICAL SERVICE LOSS DURATION			CARD TYPE	CARD NO.
	NO.	TYPE	LOCATION	CLIMATE	ATMOSPHERE	HR. PER DAY	DAYS PER WK.	PER FAILURE	PER HR. DOWNTIME			NO. OF UNITS	UNITS*			
1	4	6	8	9	10	11	13	15	20	25	31	33	36		79	80
															1	1

COLUMN	NAME	CODE	DESCRIPTION
1	Company Code		Fill in on all pages a three-letter abbreviation of company name for identification of data.
4	Plant No		Fill in on all pages a sequence number starting with "1" for Plant 1, "2" for Plant 2, etc. for identification of data. A plant may consist of one or more units at the same site.
6	Plant Type		Fill in on all pages the plant type
		1	Auto Industry
		2	Cement Industry
		3	Chemical Industry
		4	Metal Industry
		5	Mininq Industry
		6	Petroleum Industry
		7	Pulp and Paper Industry
		8	Rubber and Plastics Industry
		9	Textile Industry
		10	Other Light Manufacturing
		11	Other Heavy Manufacturing
		99	Other
8	Plant Location	1	USA and Canada
		2	Foreign
9	Plant Climate (For entire plant site)		Average of daily maximums for hottest month: Temperature / Relative Humidity (RH) (measured at noon to 2 PM ST)
		1	Hot (>90F) — High (>55 RH)
		2	Hot (>90F) — Moderate (50-55 RH)
		3	Hot (>90F) — Low (<50 RH)
		4	Moderate (80-90F) — High (>55 RH)
		5	Moderate (80-90F) — Moderate (50-55 RH)
		6	MOderate (80-90F) — Low (<50 RH)
		7	Low (<80F) — High (>55 RH)
		8	Low (<80F) — Moderate (50-55 RH)
		9	Low (<80F) — Low (<50 RH)
10	Plant Atmosphere (For entire plant site)	1	Clean to slightly polluted air
		2	With salt spray and corrosive chemicals
		3	With salt spray and dust or sand
		4	With salt spray only
		5	With corrosive chemicals and dust or sand
		6	With corrosive chemicals only
		7	With dust or sand only
		8	With conductive dust
		9	Other
	Plant Operating Schedule		
11	Hours per day		Give hours per normal working day that plant operates
13	Days per week		Give days per normal working week that plant operates
	Estimated Plant Outage Cost, Dollars		
15	Per Failure		Extra expense incurred because of a failure only (not including plant downtime), such as for damaged equipment, spoiled product, extra maintenance, or extra repair costs

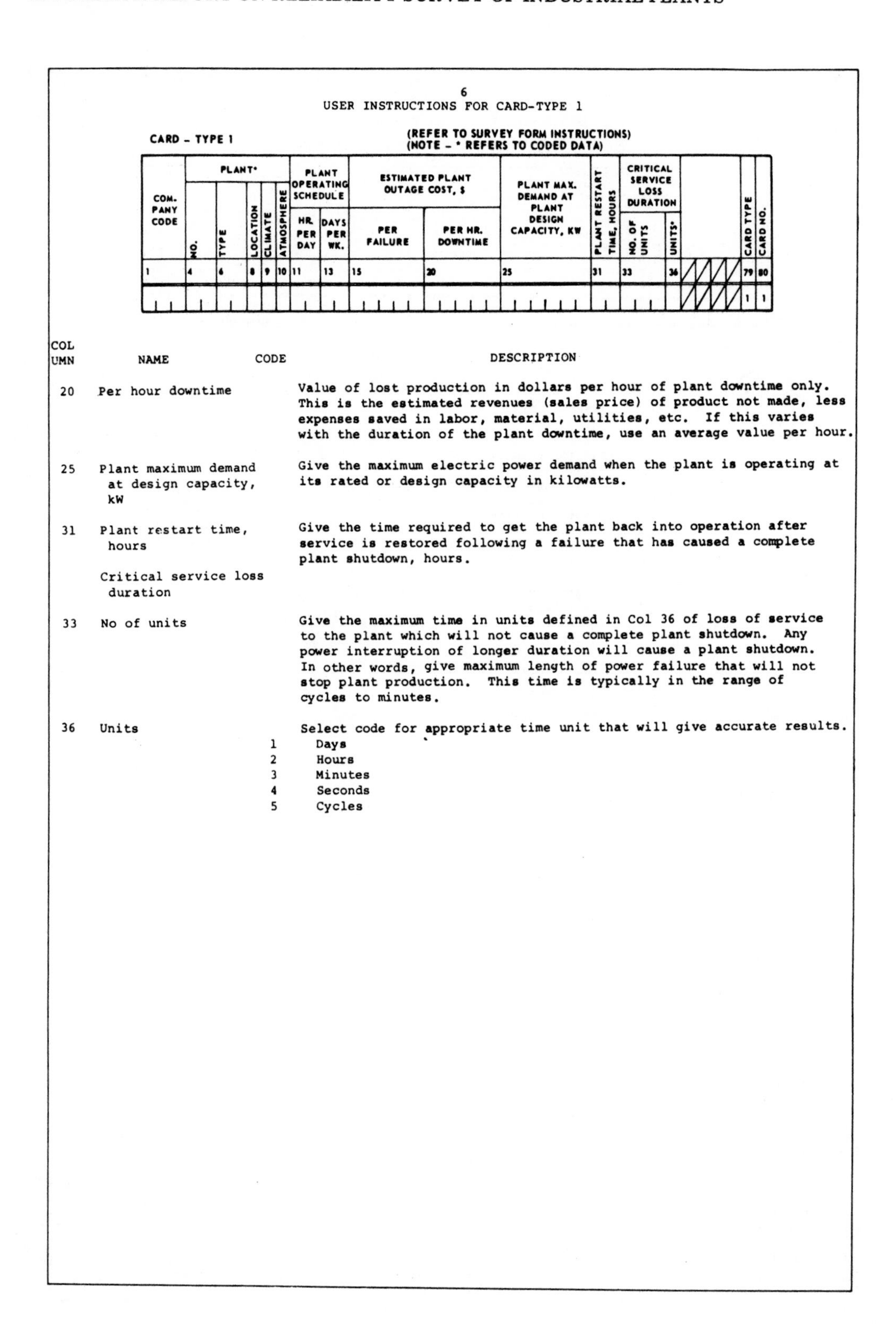

6

USER INSTRUCTIONS FOR CARD-TYPE 1

CARD - TYPE 1

(REFER TO SURVEY FORM INSTRUCTIONS)
(NOTE - * REFERS TO CODED DATA)

COMPANY CODE	PLANT*					PLANT OPERATING SCHEDULE		ESTIMATED PLANT OUTAGE COST, $		PLANT MAX. DEMAND AT PLANT DESIGN CAPACITY, KW	PLANT RESTART TIME, HOURS	CRITICAL SERVICE LOSS DURATION			CARD TYPE	CARD NO.
	NO.	TYPE	LOCATION	CLIMATE	ATMOSPHERE	HR. PER DAY	DAYS PER WK.	PER FAILURE	PER HR. DOWNTIME			NO. OF UNITS	UNITS*			
1	4	6	8	9	10	11	13	15	20	25	31	33	36		79	80
															1	1

COLUMN	NAME	CODE	DESCRIPTION
20	Per hour downtime		Value of lost production in dollars per hour of plant downtime only. This is the estimated revenues (sales price) of product not made, less expenses saved in labor, material, utilities, etc. If this varies with the duration of the plant downtime, use an average value per hour.
25	Plant maximum demand at design capacity, kW		Give the maximum electric power demand when the plant is operating at its rated or design capacity in kilowatts.
31	Plant restart time, hours		Give the time required to get the plant back into operation after service is restored following a failure that has caused a complete plant shutdown, hours.
	Critical service loss duration		
33	No of units		Give the maximum time in units defined in Col 36 of loss of service to the plant which will not cause a complete plant shutdown. Any power interruption of longer duration will cause a plant shutdown. In other words, give maximum length of power failure that will not stop plant production. This time is typically in the range of cycles to minutes.
36	Units		Select code for appropriate time unit that will give accurate results.
		1	Days
		2	Hours
		3	Minutes
		4	Seconds
		5	Cycles

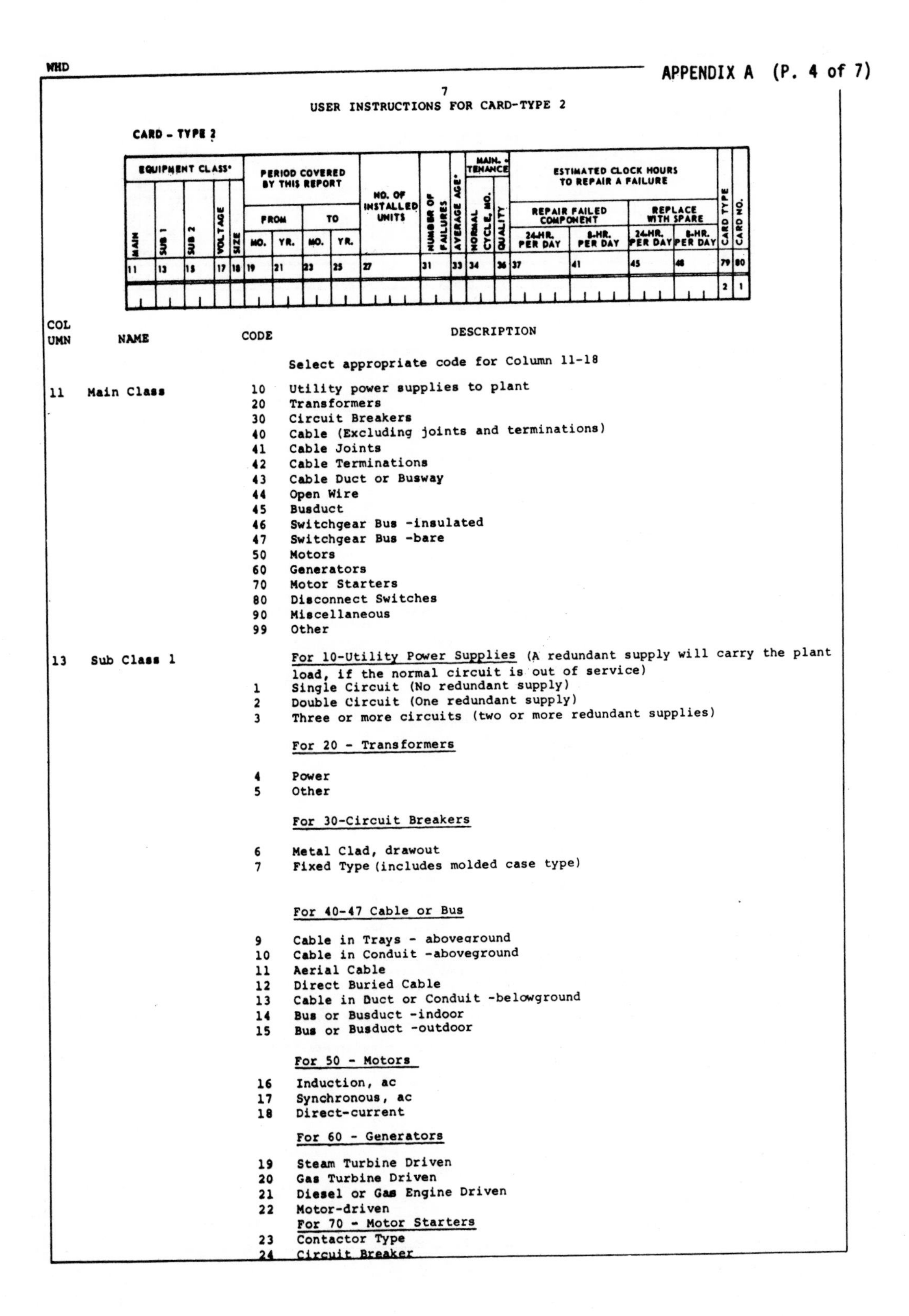

WHD

APPENDIX A (P. 4 of 7)

7

USER INSTRUCTIONS FOR CARD-TYPE 2

CARD - TYPE 2

EQUIPMENT CLASS*					PERIOD COVERED BY THIS REPORT				NO. OF INSTALLED UNITS	NUMBER OF FAILURES	AVERAGE AGE*	MAIN-TENANCE		ESTIMATED CLOCK HOURS TO REPAIR A FAILURE				CARD TYPE	CARD NO.
					FROM		TO							REPAIR FAILED COMPONENT		REPLACE WITH SPARE			
MAIN	SUB 1	SUB 2	VOLTAGE	SIZE	MO.	YR.	MO.	YR.				NORMAL CYCLE, MO.	QUALITY	24-HR. PER DAY	8-HR. PER DAY	24-HR. PER DAY	8-HR. PER DAY		
11	13	15	17	18	19	21	23	25	27	31	33	34	36	37	41	45	48	79	80
																		2	1

COLUMN	NAME	CODE	DESCRIPTION
			Select appropriate code for Column 11-18
11	Main Class	10	Utility power supplies to plant
		20	Transformers
		30	Circuit Breakers
		40	Cable (Excluding joints and terminations)
		41	Cable Joints
		42	Cable Terminations
		43	Cable Duct or Busway
		44	Open Wire
		45	Busduct
		46	Switchgear Bus -insulated
		47	Switchgear Bus -bare
		50	Motors
		60	Generators
		70	Motor Starters
		80	Disconnect Switches
		90	Miscellaneous
		99	Other
13	Sub Class 1		For 10-Utility Power Supplies (A redundant supply will carry the plant load, if the normal circuit is out of service)
		1	Single Circuit (No redundant supply)
		2	Double Circuit (One redundant supply)
		3	Three or more circuits (two or more redundant supplies)
			For 20 - Transformers
		4	Power
		5	Other
			For 30-Circuit Breakers
		6	Metal Clad, drawout
		7	Fixed Type (includes molded case type)
			For 40-47 Cable or Bus
		9	Cable in Trays - aboveground
		10	Cable in Conduit -aboveground
		11	Aerial Cable
		12	Direct Buried Cable
		13	Cable in Duct or Conduit -belowground
		14	Bus or Busduct -indoor
		15	Bus or Busduct -outdoor
			For 50 - Motors
		16	Induction, ac
		17	Synchronous, ac
		18	Direct-current
			For 60 - Generators
		19	Steam Turbine Driven
		20	Gas Turbine Driven
		21	Diesel or Gas Engine Driven
		22	Motor-driven
			For 70 - Motor Starters
		23	Contactor Type
		24	Circuit Breaker

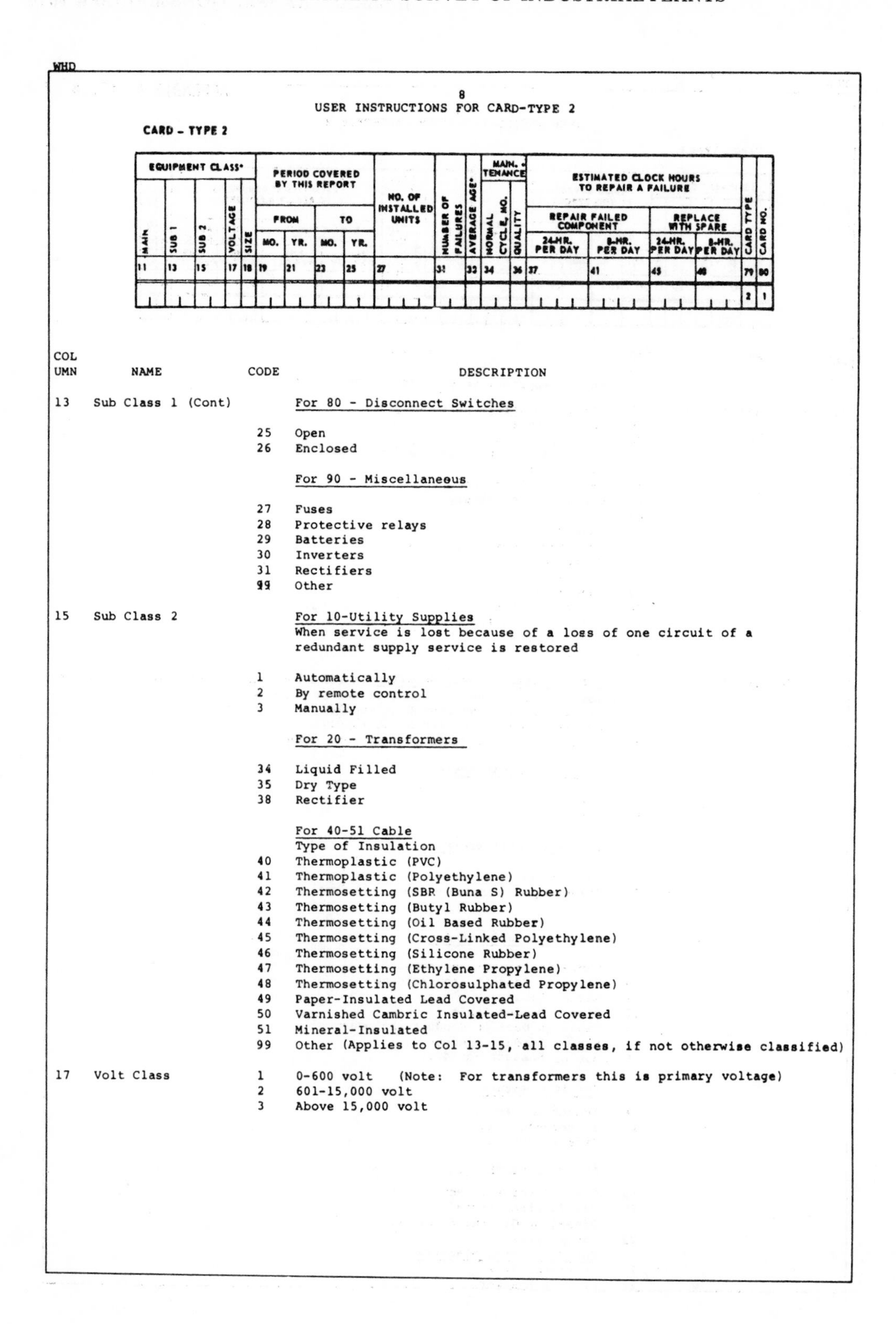

WHD

8

USER INSTRUCTIONS FOR CARD-TYPE 2

CARD - TYPE 2

EQUIPMENT CLASS*					PERIOD COVERED BY THIS REPORT				NO. OF INSTALLED UNITS	NUMBER OF FAILURES	AVERAGE AGE*	MAIN-TENANCE		ESTIMATED CLOCK HOURS TO REPAIR A FAILURE				CARD TYPE	CARD NO.
					FROM		TO							REPAIR FAILED COMPONENT		REPLACE WITH SPARE			
MAIN	SUB 1	SUB 2	VOLTAGE	SIZE	MO.	YR.	MO.	YR.				NORMAL CYCLE, MO.	QUALITY	24-HR. PER DAY	8-HR. PER DAY	24-HR. PER DAY	8-HR. PER DAY		
11	13	15	17	18	19	21	23	25	27	31	33	34	36	37	41	45	48	79	80
																		2	1

COLUMN	NAME	CODE	DESCRIPTION
13	Sub Class 1 (Cont)		For 80 - Disconnect Switches
		25	Open
		26	Enclosed
			For 90 - Miscellaneous
		27	Fuses
		28	Protective relays
		29	Batteries
		30	Inverters
		31	Rectifiers
		99	Other
15	Sub Class 2		For 10-Utility Supplies When service is lost because of a loss of one circuit of a redundant supply service is restored
		1	Automatically
		2	By remote control
		3	Manually
			For 20 - Transformers
		34	Liquid Filled
		35	Dry Type
		38	Rectifier
			For 40-51 Cable Type of Insulation
		40	Thermoplastic (PVC)
		41	Thermoplastic (Polyethylene)
		42	Thermosetting (SBR (Buna S) Rubber)
		43	Thermosetting (Butyl Rubber)
		44	Thermosetting (Oil Based Rubber)
		45	Thermosetting (Cross-Linked Polyethylene)
		46	Thermosetting (Silicone Rubber)
		47	Thermosetting (Ethylene Propylene)
		48	Thermosetting (Chlorosulphated Propylene)
		49	Paper-Insulated Lead Covered
		50	Varnished Cambric Insulated-Lead Covered
		51	Mineral-Insulated
		99	Other (Applies to Col 13-15, all classes, if not otherwise classified)
17	Volt Class	1	0-600 volt (Note: For transformers this is primary voltage)
		2	601-15,000 volt
		3	Above 15,000 volt

USER INSTRUCTION FOR CARD-TYPE 2

CARD - TYPE 2

EQUIPMENT CLASS*					PERIOD COVERED BY THIS REPORT				NO. OF INSTALLED UNITS	NUMBER OF FAILURES	AVERAGE AGE*	MAIN-TENANCE		ESTIMATED CLOCK HOURS TO REPAIR A FAILURE				CARD TYPE	CARD NO.
					FROM		TO							REPAIR FAILED COMPONENT		REPLACE WITH SPARE			
MAIN	SUB 1	SUB 2	VOLTAGE	SIZE	MO.	YR.	MO.	YR.				NORMAL CYCLE, MO.	QUALITY	24-HR. PER DAY	8-HR. PER DAY	24-HR. PER DAY	8-HR. PER DAY		
11	13	15	17	18	19	21	23	25	27	31	33	34	36	37	41	45	48	79	80
																		2	1

COLUMN	NAME	CODE	DESCRIPTION
18	Size Class		For Main Class 10 - Utility Supplies For Main Class 30 - Circuit Breakers For Main Class 80 - Disc Switches For Main Class 90 - Miscellaneous, Fuses
		1	100-600 amperes
		2	Above 600 amperes
			For Main Class 20 - Transformers
		3	300-750 kVA
		4	751-2499 kVA
		5	2500-up kVA
			For Main Class 40-45 - Cable, etc
		6	Above No 1 AWG
			For Main Class 50 - Motors For Main Class 70 - Motor Starters
		7	50-1500 horsepower
		8	Above 1500 horsepower
			For Main Class 60 - Generators
		9	500-up kW
	Period covered by this report		Give month and year (numerals) for period for which failure data is available
19	From: Mo		Starting Month (Try to include data from date of installation)
21	From: Yr		Starting Year
23	To: Mo		Ending Month (Try to include data to date of this report)
25	To: Yr		Ending Year
27	No of installed units		Give total number of units installed. For cable or open wire, give length of circuit or run in M ft. For cable duct or busduct, give circuit length in feet. For switchgear bus, give the number of connected circuit breakers or instrument transformer compartments. For utility power supplies, give the number of separate supplies.
31	No of Failures		Give total number of failures that occurred during period of report. If more than 10 use additional page.
			Select codes for Column 30-53
33	Average Age	1	Less than 1 year old
		2	1-10 years old
		3	More than 10 years old
34	Maintenance Normal Cycle, Mo		Give normal cycle for preventive maintenance - (even if a failure has not occurred)
		1	Less than 12 months
		2	12-24 months
		3	More than 24 months
		4	No preventive maintenance
36	Maintenance Quality		Your estimate of quality of preventive maintenance is -
		1	Excellent (by own forces)
		2	Fair (by own forces)
		3	Poor, inadequate (by own forces)
		4	None
		5	Excellent (by contracted forces)
		6	Fair (by contracted forces)
		7	Poor inadequate (by contracted forces)

WHD

10

USER INSTRUCTIONS FOR CARD-TYPE 2

CARD - TYPE 2

EQUIPMENT CLASS*					PERIOD COVERED BY THIS REPORT				NO. OF INSTALLED UNITS	NUMBER OF FAILURES	AVERAGE AGE*	MAIN-TENANCE		ESTIMATED CLOCK HOURS TO REPAIR A FAILURE				CARD TYPE	CARD NO.
					FROM		TO							REPAIR FAILED COMPONENT		REPLACE WITH SPARE			
MAIN	SUB 1	SUB 2	VOLTAGE	SIZE	MO.	YR.	MO.	YR.				NORMAL CYCLE, MO.	QUALITY	24-HR. PER DAY	8-HR. PER DAY	24-HR. PER DAY	8-HR. PER DAY		
11	13	15	17	18	19	21	23	25	27	31	33	34	36	37	41	45	48	79	80
																		2	1

COLUMN	NAME	CODE	DESCRIPTION
	Estimated clock hours		Repair time (see definitions) Fill in the clock time for diagnosing the trouble, locating the failed component, waiting for parts repairing or replacing, testing and restoring the component to service. This is your estimate of the average repair time. Please note that actual repair times are requested in CARD-TYPE 3, Col 26. Explain on reverse side how work is done if by other than own forces.
	Repair failed component		With repair of failed equipment
37	24-hr per day		On round-the-clock emergency basis
41	8-hr per day		On basis of repair during normal work day
			With replacement of failed equipment with a spare by removal of failed equipment and substitution of spare equipment
	Repair with spare		
45	24-hr per day		On round-the-clock emergency basis
48	8-hr per day		On basis of repair during normal work day

USER INSTRUCTIONS FOR CARD-TYPE 3

CARDS – TYPE 3

FAILURE															LOADS LOST*					% PRODUCTION LOST*	PLANT OUTAGE DURATION		SERVICE RESTORED*	CARD TYPE	CARD NO.
NUMBER	DATE		FOREWARNING*	DURATION		REPAIR METHOD*	REPAIR URGENCY*	NO. SINCE LAST MAINTAINED*	DAMAGED PART*	TYPE*	RESPONSIBILITY*	INITIATING CAUSE*	CONTRIBUTING CAUSE*	CHARACTERISTICS*	COMPUTER	MOTOR	LIGHTING	SOLENOID	OTHER		NO. OF UNITS	UNITS*			
	MO.	YR.		NO. OF UNITS	UNITS*																				
19	21	23	23	26	29	30	32	34	36	38	40	42	44	46	48	49	50	51	52	53	54	57	58	79	80
1																								3	1

COLUMN	NAME	CODE	DESCRIPTION
19	Failure No		Fill in one card (line) for each failure. The last failure number in Col 19 should correspond with the total failures reported in Col 31 of CARD-TYPE 2. If that number was "0" then no TYPE 3 cards should be filled in.
	Failure Date		
21	Mo		Fill in month failure occured (numeral)
23	Yr		Fill in year failure occurred (numeral)
25	Failure Forewarning		For public utility power interruption only
		1	If no forewarning was given
		2	If forewarning was given
			For other types of failure, leave blank
	Failure Duration		Fill in duration of failure from its initiation until (1) service is restored to normal, if a power interruption, or (2) the affected component or its replacement once again becomes available to perform its intended function.
26	No of Units		Fill in the number of time units selected in Col 29.
29	Units		Select code for appropriate time unit that will give accurate results. For most cases select hours as unit.
		1	Days
		2	Hours
		3	Minutes
		4	Seconds
		5	Cycles
			Select code for Col 30-44 (Leave blank for utility failures)
30	Failure Repair Method	1	Repair of failed component in place or sent out for repair
		2	Repair by replacement of failed component with spare
32	Failure Repair Urgency	1	Requiring round-the-clock all out efforts
		2	Requiring repair work only during regular workday, perhaps with some overtime.
		3	Requiring repair work on a non-priority basis.
34	Failure, months since maintained		Failed component last had preventive maintenance -
		1	Less than 12 months ago
		2	12-24 months ago
		3	Over 24 months ago
		4	No preventive maintenance
36	Failure, Damaged Part	1	Insulation - winding
		2	Insulation - bushing
		3	Insulation - other
		4	Mechanical - bearings
		5	Mechanical - other moving parts
		6	Mechanical - other
		7	Other electrical - auxiliary device
		8	Other electrical - protective device
		9	Tap changer - no load type
		10	Tap changer - load type
		99	Other

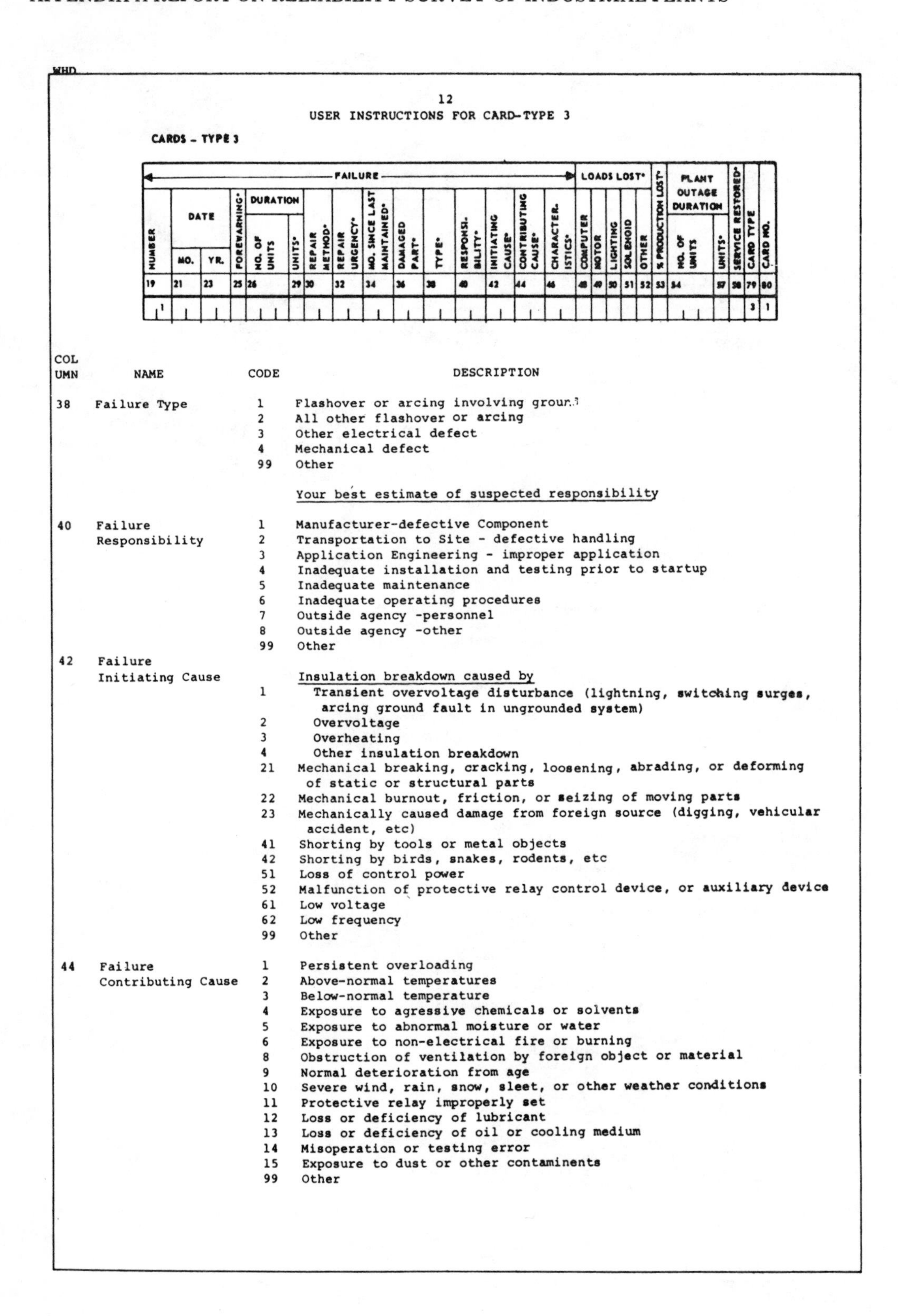

WHD

12

USER INSTRUCTIONS FOR CARD-TYPE 3

CARDS - TYPE 3

FAILURE															LOADS LOST*					% PRODUCTION LOST*	PLANT OUTAGE DURATION		SERVICE RESTORED*	CARD TYPE	CARD NO.
NUMBER	DATE		FOREWARNING*	DURATION		REPAIR METHOD*	REPAIR URGENCY*	MO. SINCE LAST MAINTAINED*	DAMAGED PART*	TYPE*	RESPONSIBILITY*	INITIATING CAUSE*	CONTRIBUTING CAUSE*	CHARACTERISTICS*	COMPUTER	MOTOR	LIGHTING	SOLENOID	OTHER		NO. OF UNITS	UNITS*			
	MO.	YR.		NO. OF UNITS	UNITS*																				
19	21	23	25	26	29	30	32	34	36	38	40	42	44	46	48	49	50	51	52	53	54	57	58	79	80
1																								3	1

COLUMN	NAME	CODE	DESCRIPTION
38	Failure Type	1	Flashover or arcing involving ground
		2	All other flashover or arcing
		3	Other electrical defect
		4	Mechanical defect
		99	Other
			Your best estimate of suspected responsibility
40	Failure Responsibility	1	Manufacturer-defective Component
		2	Transportation to Site - defective handling
		3	Application Engineering - improper application
		4	Inadequate installation and testing prior to startup
		5	Inadequate maintenance
		6	Inadequate operating procedures
		7	Outside agency -personnel
		8	Outside agency -other
		99	Other
42	Failure Initiating Cause		Insulation breakdown caused by
		1	Transient overvoltage disturbance (lightning, switching surges, arcing ground fault in ungrounded system)
		2	Overvoltage
		3	Overheating
		4	Other insulation breakdown
		21	Mechanical breaking, cracking, loosening, abrading, or deforming of static or structural parts
		22	Mechanical burnout, friction, or seizing of moving parts
		23	Mechanically caused damage from foreign source (digging, vehicular accident, etc)
		41	Shorting by tools or metal objects
		42	Shorting by birds, snakes, rodents, etc
		51	Loss of control power
		52	Malfunction of protective relay control device, or auxiliary device
		61	Low voltage
		62	Low frequency
		99	Other
44	Failure Contributing Cause	1	Persistent overloading
		2	Above-normal temperatures
		3	Below-normal temperature
		4	Exposure to agressive chemicals or solvents
		5	Exposure to abnormal moisture or water
		6	Exposure to non-electrical fire or burning
		8	Obstruction of ventilation by foreign object or material
		9	Normal deterioration from age
		10	Severe wind, rain, snow, sleet, or other weather conditions
		11	Protective relay improperly set
		12	Loss or deficiency of lubricant
		13	Loss or deficiency of oil or cooling medium
		14	Misoperation or testing error
		15	Exposure to dust or other contaminents
		99	Other

USER INSTRUCTIONS FOR CARD-TYPE 3

CARDS – TYPE 3

FAILURE															LOADS LOST*					% PRODUCTION LOST*	PLANT OUTAGE DURATION		SERVICE RESTORED*	CARD TYPE	CARD NO.
NUMBER	DATE		FOREWARNING*	DURATION		REPAIR METHOD*	REPAIR URGENCY*	NO. SINCE LAST MAINTAINED*	DAMAGED PART*	TYPE*	RESPONSIBILITY*	INITIATING CAUSE*	CONTRIBUTING CAUSE*	CHARACTERISTICS*	COMPUTER	MOTOR	LIGHTING	SOLENOID	OTHER		NO. OF UNITS	UNITS*			
	MO.	YR.		NO. OF UNITS	UNITS*																				
19	21	23	25	26	29	30	32	34	36	38	40	42	44	46	48	49	50	51	52	53	54	57	58	79	80
1																								3	1

COLUMN	NAME	CODE	DESCRIPTION
46	Failure Characteristic		Utility Power Supplies (Select code)
		1	Failure of single circuit (No redundant supply)
		2	Failure of one circuit of a double-circuit redundant supply
		3	Failure of both circuits of a double-circuit redundant supply
		4	Failure of all circuits of a three or more circuit redundant supply
		5	Partial failure of a three or more circuit redundant supply
			Transformers (Select code)
		6	Automatic removal by protective equipment
		7	Partial failure reducing capacity
		8	Manual removal
			Circuit Breakers (Select code)
		9	Failed to close when it should
		10	Failed while opening
		11	Opened when it shouldn't
		12	Damaged while successfully opening
		13	Damaged while closing
		14	Failed while operating (not while opening or closing)
			General (Select code for any other class)
		15	Failed (this applies to all classes)
		16	Failed during testing or maintenance
		17	Damage discovered during testing or maintenance
		20	Partial failure
		99	Other
	Loads Lost		What loads were lost because of failure (1=yes, 0=no, 9= not known) even though power is restored promptly
48	Computer		One or more computers or solid-state control devices operated incorrectly
49	Motor		One or more motors (contactor dropout)
50	Lighting		Lighting load
51	Solenoid		One or more solenoid -operated devices dropped out, such as a solenoid-operated fuel valve
52	Other		Lost other loads, describe in remarks
53	Percent Production Lost	0	None
		1	0-30 percent
		2	Above 30 percent

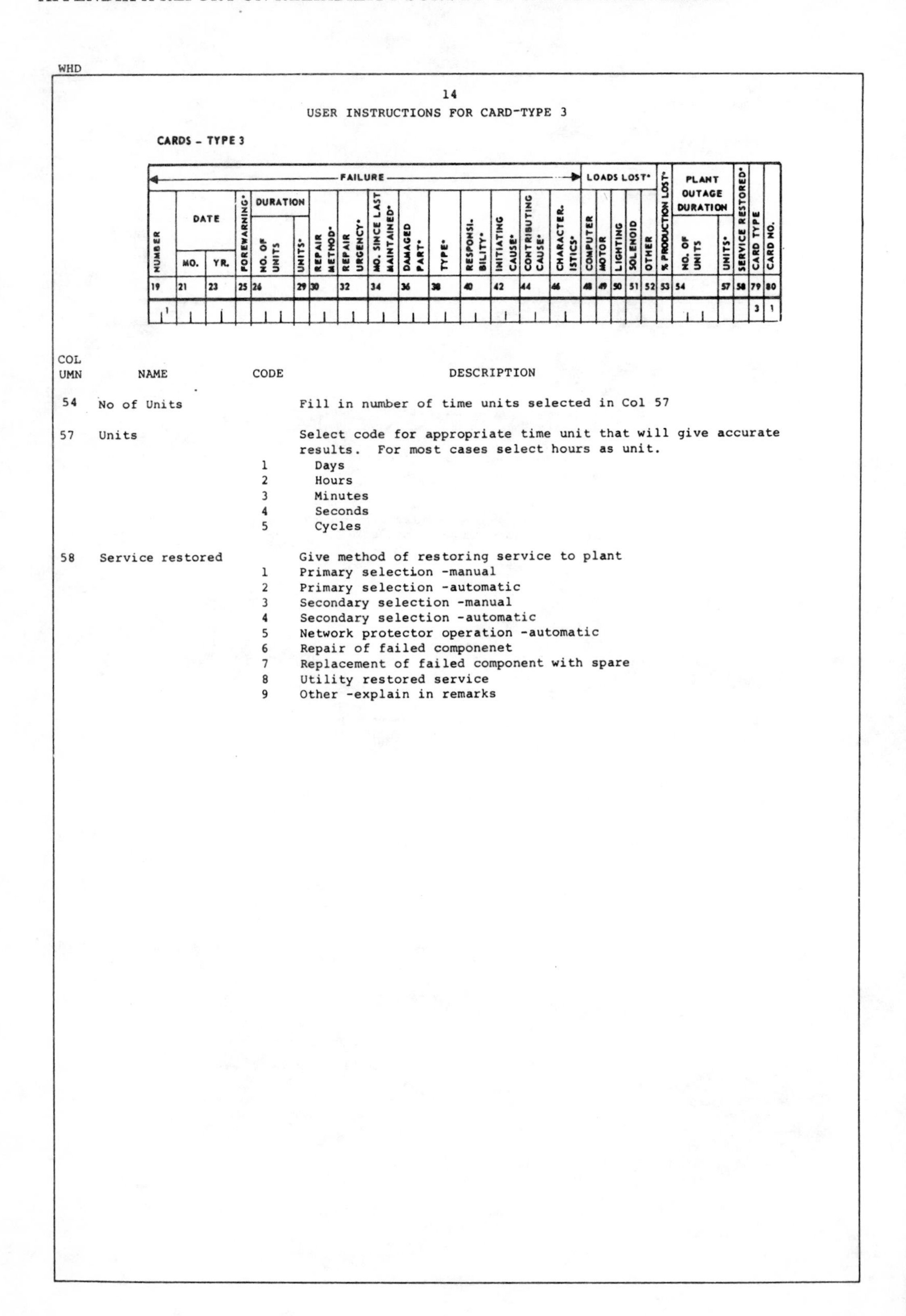

WHD

14

USER INSTRUCTIONS FOR CARD-TYPE 3

CARDS - TYPE 3

FAILURE															LOADS LOST*					% PRODUCTION LOST*	PLANT OUTAGE DURATION		SERVICE RESTORED*	CARD TYPE	CARD NO.
NUMBER	DATE		FOREWARNING*	DURATION		REPAIR METHOD*	REPAIR URGENCY*	NO. SINCE LAST MAINTAINED*	DAMAGED PART*	TYPE*	RESPONSI. BILITY*	INITIATING CAUSE*	CONTRIBUTING CAUSE*	CHARACTER. ISTICS*	COMPUTER	MOTOR	LIGHTING	SOLENOID	OTHER						
	MO.	YR.		NO. OF UNITS	UNITS*																NO. OF UNITS	UNITS*			
19	21	23	25	26	29	30	32	34	36	38	40	42	44	46	48	49	50	51	52	53	54	57	58	79	80
1																								3	1

COLUMN	NAME	CODE	DESCRIPTION
54	No of Units		Fill in number of time units selected in Col 57
57	Units		Select code for appropriate time unit that will give accurate results. For most cases select hours as unit.
		1	Days
		2	Hours
		3	Minutes
		4	Seconds
		5	Cycles
58	Service restored		Give method of restoring service to plant
		1	Primary selection -manual
		2	Primary selection -automatic
		3	Secondary selection -manual
		4	Secondary selection -automatic
		5	Network protector operation -automatic
		6	Repair of failed componenet
		7	Replacement of failed component with spare
		8	Utility restored service
		9	Other -explain in remarks

Discussion

Motors

The data in Tables 7 and 2 show that synchronous motors, 0–600 V, have a failure rate approximately 15 times lower than induction motors, 0–600 V. It is believed that the failure 0.0007 per year for synchronous motors, 0–600 V, is much too low and is in error. It is believed that synchronous and induction motors, 0–600 V, should have failure rates that are nearly the same.

Generators

The data in Tables 8 and 2 show that steam turbine driven generators have a failure rate almost 20 times lower than gas turbine driven generators. It is believed that the failure rate of 0.032 per year for steam turbine driven generators is too low; the failure rate should probably be several times higher than this value. The gas turbine data in Table 8 show that one plant in the petroleum industry had 54 failures in 5.5 unit-years; this compares with 3 failures in 83.9 unit-years for the other three plants that submitted data in the survey. It is believed that the overall failure rate of 0.638 per year for gas turbines is too high.

Open Wire

A clear definition was not given for "open wire" on the survey form (see Appendix A). It is believed that all of the respondents interpreted "open wire" to mean "bare or weatherproof conductors supported on insulators."

Cable

The data in Tables 13 and 2 show that cable above ground and aerial has a failure rate for 0–600 V that is ten times lower than 601-15 000 V. It is believed that the failure rate of 0.00141 per unit-year for 0–600 V above ground and aerial is too low.

There is a wide variation in the failure rate for cable, 601-15 000 V, based upon the application (in trays above ground, in conduit above ground, aerial cable, in duct or conduit below ground). This variation covers a range of 8 to 1. It is believed that the failure rate of 0.04918 per year is too high for cable, 601-15 000 V, in conduit above ground.

There is a wide variation in the cable failure rate shown in Table 14 (and Table 2) for the different types of insulation (601-15 000 V, all applications). These failure rates vary over a range of 5 to 1. The very low failure rate data for thermoplastic insulation and the high failure rate data for other insulation came primarily from the chemical industry.

Switchgear Bus

The failure rate in Table 10 (and Table 2) shows that insulated bus, 601-15 000 V, has a failure rate about three times higher than bare bus, above 600 V. It is believed that this is the opposite of what it should be. The data submitted by the chemical industry has caused this distortion; they had a very high failure rate for insulated bus (601-15 000 V) and a low failure rate for bare bus (above 600 V).

Electric Utility Power Supplies

The data for electric utility power supplies are shown in Tables 3 and 2. The failure rate is about the same for a single circuit and a double or triple circuit. This is evidently due to the predominance of the throwover mode of operation of multiple-circuit supplies. However, the actual downtime per failure is about three to nine times higher for a single circuit than for a double or triple circuit; the downtime depends on whether manual switchover or automatic switchover is used on a multiple-circuit system.

It appears that many respondents misinterpreted the "number of installed units" for double- or triple-circuit electric utility power supplies. What was desired was the number of separate and independent points of supply, but this was often interpreted to be the number of circuits in the utility supply system. Thus the tendency was to report two installed units for double-circuit supplies. It is believed that this error was made in almost every case. Therefore, *the Reliability Subcommittee changed the number of installed units for multiple-circuit utility supplies to 1 except in those cases where other evidence indicated the presence of more than one point of supply.* The sample size shown in Tables 3 and 2 reflects this change for double- or triple-circuit electric utility power supplies. Thus a double- or triple-circuit supply for one year is counted as one unit-year.

It also appears that a few respondents incorrectly interpreted failure duration on card type 3 for multiple-circuit electric utility supplies. What was desired was the period of time during which service was interrupted. However, in a few cases it appears that what was given was the time to repair one circuit of a multiple-circuit supply even though the supply interruption time is limited to the time required to throw over to the alternate supply circuit. The *Reliability Subcommittee changed the failure duration to the value given for plant outage duration in those cases in which such an error was believed to exist.* However, it is suspected that not all of these errors were corrected. The effect of this change was to reduce the actual hours of downtime per failure for multiple-circuit supplies. The majority of the multiple-circuit supply failures are due to loss of the normal feed, and the duration of the failure is limited to the time to switch to the alternate feed. The average outage duration in Tables 3 and 2 is shorter for automatic switching than for manual switching, as one would expect.

There were 25 recorded cases of simultaneous failure of all circuits in a double- or triple-circuit supply. This gives a failure rate of 0.119 failure per year for loss of all circuits at one time. Further details on this are given in Part 3 [13]. Thus a multiple-circuit electric utility power supply has a failure rate (loss of all circuits at one time) that is only about five times lower than the failure rate (0.537 failures per year) for a single-circuit supply and about six times lower than the all-inclusive failure rate of 0.643 failure per year. The ratio between all-inclusive failure rate and the failure rate for loss of all circuits at one time is not as large as one might suspect. Some of the reasons for this are the following.

1) Some portion of utility supply failures are due to failure of the bulk power system which feeds all the supply circuits.

2) At least some cases of loss of all circuits at one time occur when a forced outage of one circuit overlaps a scheduled or maintenance outage of the other circuit (typical utility industry data indicate that this type of overlapping outage is often more probable than overlapping forced outages).

3) The all-inclusive failure rate is, in effect, an average outage rate reflecting the performance of some throwover schemes and some normally closed breaker schemes. Thus, since throw-

over schemes are expected to have higher outage rates than normally closed breaker schemes, it follows that the computed all-inclusive outage rate is probably somewhat lower than the outage rate which would be computed for throwover schemes only. (Unfortunately we cannot compute the throwover scheme outage rate since we do not know which of the reported utility supplies are throwover schemes.)

Only point 3) reflects on the accuracy of the data; the other two points just reflect the facts of life.

A comparison of the all-inclusive failure rate (0.643 failures per year) with the failure rate for loss of all circuits at one time (0.119 failures per year) gives a rough idea of the degree of supply failure rate improvement possible by going from a throwover scheme to a scheme using normally closed circuit breakers.

References

[1] W. H. Dickinson, P. E. Gannon, C. R. Heising, A. D. Patton, and D. W. McWilliams, "Fundamentals of reliability techniques as applied to industrial power systems," in *Conf. Rec. 1971 IEEE Ind. Comm. Power Syst. Tech. Conf.* 71C18-IGA, pp. 10–31.

[2] C. R. Heising, "Reliability and availability comparison of common low-voltage industrial power distribution systems," *IEEE Trans. Ind. Gen. Appl.*, vol. IGA-6, pp. 416–424, Sept./Oct. 1970.

[3] W. H. Dickinson, "Economic evaluation of industrial power system reliability," *AIEE Trans. (Appl. Ind.)*, vol. 76, pp. 264–271, Nov. 1957.

[4] W. H. Dickinson, "Evaluation of alternative power distribution systems for refinery process units," *AIEE Trans. (Power Appl. Syst.)*, vol. 79, Apr. 1960.

[5] W. H. Dickinson, "Economic justification of petroleum industry automation and other alternatives by the revenue requirements method," *IEEE Trans. Ind. Gen. Appl.*, vol. IGA-1, pp. 39–50, Jan./Feb. 1965.

[6] D. P. Garver, F. E. Montmeat, and A. D. Patton, "Power systems reliability I–Measures of reliability and methods of calculation," *IEEE Trans. Power Appl Sys.*, vol. 83, pp. 727–737, July 1964.

[7] F. E. Montmeat, A. D. Patton, J. Zemkoski, and D. J. Cummin, "Power system reliability II–Applications and a computer program," *ibid.*, vol. PAS-84, pp. 636–643, July 1965.

[8] Z. G. Todd, "A probability method for transmission and distribution outage calculations," *IEEE Trans. Power App. Syst.*, vol. 83, pp. 695–701, July 1964.

[9] C. F. DeSieno, and L. L. Stine, "A probability method for determining the reliability of electric power systems," *IEEE Trans. Power App. Syst.*, vol. 83, pp. 174–181, Feb. 1964.

[10] "General principles for reliability analysis of nuclear power generating station protection systems," IEEE Publ. 352, ANSI N41.4, 1972.

[11] W. H. Dickinson, "Report on reliability of electric equipment in industrial plants," *AIEE Trans. (Appl. Ind.)*, vol. 81, pp. 132–151, July 1962.

[12] IEEE Committee Report, "Report on reliability survey of industrial plants, Part II: Cost of power outages, plant restart time, critical service loss duration time, and type of loads lost versus time of power outages," this issue, pp. 236–241.

[13] IEEE Committee Report, "Report on reliability survey of industrial plants; Part III: Causes and types of failures of electrical equipment, methods of repair, and urgency of repair," this issue, pp. 242–249.

[14] N. H. Roberts, *Mathematical Methods in Reliability Engineering.* New York: McGraw-Hill, 1964.

[15] I. Bazovsky, *Reliability Theory and Practice.* Englewood Cliffs, N.J.: Prentice-Hall, 1961.

[16] D. K. Lloyd and M. Lipow, *Reliability: Management, Methods, and Mathematics.* Englewood Cliffs, N.J.: Prentice-Hall, 1962.

[17] A. D. Patton, "Determination and analysis of data for reliability studies," *IEEE Trans. Power App. Syst.*, vol. PAS-87, pp. 84–100, Jan. 1968.

Report on Reliability Survey of Industrial Plants, Part II: Cost of Power Outages, Plant Restart Time, Critical Service Loss Duration Time, and Type of Loads Lost Versus Time of Power Outages

IEEE COMMITTEE REPORT

***Abstract*—An IEEE sponsored reliability survey of industrial plants was completed during 1972. This survey included the cost of power outages, plant restart time, critical service loss duration time, and type of loads lost versus power outage duration time. Survey results reflect data from 30 companies covering 68 plants in nine industries in the United States and Canada. This information is useful in the design of industrial power distribution systems.**

Introduction

KNOWLEDGE of the cost of power outages and of plant restart time is important information for use in the design of industrial power distribution systems. In addition it is also desirable to know the critical service loss duration time and the type of loads lost versus the time of power outage.

During 1972 the Reliability Subcommittee of the IEEE Industrial and Commercial Power Systems Committee completed a reliability survey of industrial plants. This is the second part, which reports results from the survey. Included in this paper are the following results:

1) cost of power outages to industrial plants in the United States and Canada (dollars per kilowatt interrupted plus dollars per kilowatthour of undelivered energy);
2) plant restart time after a failure that has caused complete plant shutdown;
3) critical service loss duration time, that is, the maximum length of power failure that will not stop plant production;
4) type of loads lost versus the time of power outage (this includes computer, motor, lighting, and solenoid loads, and gives plant outage duration times resulting from these failures).

Paper TOD-73-158, approved by the Industrial and Commercial Power Systems Committee of the IEEE Industry Applications Society for presentation at the 1973 Industrial and Commercial Power Systems Technical Conference, Atlanta, Ga., May 13–16. Manuscript released for publication November 5, 1973.

Members of the Reliability Subcommittee of the IEEE Industrial and Commercial Power Systems Committee are W. H. Dickinson, *Chairman*, P. E. Gannon, M. D. Harris, C. R. Heising, D. W. McWilliams, R. W. Parision, A. D. Patton, and W. J. Pearce.

Survey Form

The survey form used is shown in Appendix A of Part 1 [1]. The information on the cost of power outages came from card type 1, columns 13, 20, and 25. Card type 1 also contained plant restart time (column 31) and critical service loss duration (columns 33 and 36).

The data on type of loads lost came from card type 3, columns 48, 49, 50, 51, and 52. The data on time of power outage came from columns 26 and 29 of card type 3; these data are actually the outage duration time after a failure of the electric utility power supply or a failure of electrical equipment in the power distribution system.

Response to Survey

A total of 30 companies responded to the survey questionnaire reporting data on 68 plants from nine industries in the United States and Canada. Every response did not supply all the information requested on every question. Tables 22–29 give data on how many plants provided answers to the various questions.

Statistical Analysis

The results were compiled for the United States and Canada. Data from one foreign plant are also included separately.

Survey Results

Cost of Power Outages

Each plant was asked to report data on the cost of power outages as follows:

1) Dollars per failure, i.e., extra expense incurred because of a failure only (not including plant downtime) such as for damaged equipment, spoiled product, extra maintenance, or extra repair costs.

2) Dollars per hour of downtime, i.e., value of lost production in dollars per hour of plant downtime only. This is the estimated revenues (sales price) of product not made, less expenses saved in labor, material, utilities, etc. If this varies with the duration of the plant downtime, an average value per hour was to be given.

3) Maximum electric power demand when the plant is operating at its rated or design capacity in kilowatts.

This made it possible to calculate an estimate of the cost of power outages in terms of the dollars per kilowatts interrupted plus the dollars per kilowatthours of undelivered energy. The average cost of power outages from the survey is given in Table 20.

Of the 41 plants that reported outage cost data in the survey, 31 had a maximum demand greater than 1000 kW and 10 had a maximum demand less than 1000 kW. Cost data for plants with maximum demands less than 1000 kW are not considered particularly reliable due to the small number of such plants represented in the data.

There is a wide spread in the cost of power outages. Consequently few plants with high outage costs can have a significant effect on the overall average cost. In such cases the median cost of power outages may be more representative than the average cost. The median cost is such that half of the plants have a cost greater than this value and half have less. Table 21 shows the median power outage costs. Additional details on the cost of power outages are given in Tables 22-27. These additional details include: 1) number of plants reporting the outage cost per failure and the outage cost per hour of downtime, 2) minimum plant cost, 3) maximum plant cost, 4) costs for various industries.

Tables 22, 24, and 26 give the cost of outage per failure per kilowatt maximum demand. Tables 23, 25, and 27 give the cost of a sustained outage per hour down per kilowatt maximum demand.

Plant Restart Time

Each plant was asked to report data on the time required to get the plant back into operation after service is restored following a failure that has caused a complete plant shutdown. A total of 43 plants reported these data. The average plant

TABLE 20 - AVERAGE COST OF POWER OUTAGES FOR INDUSTRIAL PLANTS IN THE UNITED STATES OF AMERICA AND CANADA

All Plants	\$1.89 per kW + \$2.68 per kWh
Plants > 1000 kW Max. Demand	\$1.05 per kW + \$0.94 per kWh
Plants < 1000 kW Max. Demand	\$4.59 per kW + \$8.11 per kWh

TABLE 21 - MEDIAN COST OF POWER OUTAGES FOR INDUSTRIAL PLANTS IN THE UNITED STATES OF AMERICA AND CANADA

All plants	\$0.69 per kW + \$0.83 per kWh
Plants > 1000 kW Max. Demand	\$0.32 per kW + \$0.36 per kWh
Plants < 1000 kW Max. Demand	\$3.68 per kW + \$4.42 per kWh

restart time was 17 h. The median was 4 h. Additional details are given in Table 28.

Critical Service Loss Duration Time

One of the most commonly asked questions is, What is a power failure? In particular, How long can power be lost without causing a complete plant shutdown? Each plant was asked to report data giving the maximum length of power failure that will not stop plant production. This time is typically in the range of cycles to minutes and is called "critical service loss duration time."

A total of 55 plants reported data on critical service loss duration time. The median value was 10 s, that is, half of the plants were greater than this value and half were less. Additional details are given in Table 29.

Loads Lost Versus Time of Power Outage

Each plant was asked, What loads were lost because of failure even though power was restored promptly? Five types of loads were included in the survey:

1) *computer:* one or more computers or solid-state control devices operated incorrectly;
2) *motor:* one or more motors (contactor dropout);
3) *lighting:* lighting load;
4) *solenoid:* one or more solenoid-operated devices dropped out, such as a solenoid-operated fuel valve;
5) *other:* lost other loads, to be described in remarks.

A very short outage duration time after an equipment failure (including electric utility power supply) might not result in a loss of load. Table 30 shows how short power outage duration times after an equipment failure affected the loads lost. The average plant outage duration resulting from these failures is also given in Table 30.

Discussion of Results

Cost of Power Outages (Tables 20–27)

1) There is a wide spread in the cost of power outages (per kilowatt and per kilowatthour) of industrial plants. Even within a given industry, such as chemical, there is a wide spread in the cost of power outages (per kilowatt and per kilowatthour) for different plants.

2) Plants with a maximum demand of less than 1000 kW have a much higher cost of power outages (per kilowatt and per kilowatthour) than plants with a maximum demand of greater than 1000 kW. This indicates that small industrial plants have a higher cost of power outages (per kilowatt and per kilowatthour) than large industrial plants. It is suspected that this may be because the small industrial plants have more employees per kilowatt (and per kilowatthour). It is also possible that high-consumption industries tend to have a lot of electrochemical or heating processes, and these tend to have low outage costs; for example, heat not supplied now can be supplied later, providing the outage is not too long.

3) It is suggested that the "all-industry" data for the 41 and 42 plants should be compiled to show 25 percent and 75 percent in addition to the minimum median and maximum values already tabulated (Tables 22 and 23).

4) It is suggested that future surveys also include the cost of power outages (per kilowatt and per kilowatthour) of commercial buildings.

TABLE 22 - PLANT OUTAGE COST PER FAILURE PER kW OF MAXIMUM DEMAND - ALL PLANTS ($ per kW)

Industry	Number of Plants Reporting	Minimum	Median	Maximum	Average
All Industry - USA & Canada	42	.002	.69	10.00	1.89
Auto	0	-	-	-	-
Cement	0	-	-	-	-
Chemical	11	.02	.22	3.33	.75
Metal	2	.18	2.42	4.67	2.42
Mining	0	-	-	-	-
Petroleum	5	.002	.07	.31	.12
Pulp and Paper	1	.33	.33	.33	.33
Rubber and Plastics	2	.28	.50	.71	.50
Textile	2	.07	1.00	1.92	1.00
Other Light Manufacturing	6	.09	1.10	2.80	1.22
Other Heavy Manufacturing	8	1.67	3.85	10.00	5.11
Other	5	.25	.94	7.50	2.86
Foreign	1	.33	.33	.33	.33

TABLE 23 - PLANT OUTAGE COST PER HR. DOWNTIME PER kW OF MAXIMUM DEMAND - ALL PLANTS ($ per kWh)

Industry	Number of Plants Reporting	Minimum	Median	Maximum	Average
All Industry - USA & Canada	41	.0009	.83	27.00	2.68
Auto	0	-	-	-	-
Cement	0	-	-	-	-
Chemical	12	.0009	.14	2.11	.33
Metal	2	.55	.94	1.33	.94
Mining	0	-	-	-	-
Petroleum	2	.04	1.24	2.43	1.24
Pulp and Paper	1	.07	.07	.07	.07
Rubber and Plastics	3	.28	.36	1.33	.66
Textile	1	.24	.24	.24	.24
Other Light Manufacturing	6	.33	.79	2.00	.91
Other Heavy Manufacturing	8	.93	6.35	27.00	9.73
Other	6	.75	2.50	5.77	2.69
Foreign	1	.07	.07	.07	.07

TABLE 24 - PLANT OUTAGE COST PER FAILURE PER kW OF MAXIMUM DEMAND - PLANTS MORE THAN 1,000 kW MAX. DEMAND ($ per kW)

Industry	Number of Plants Reporting	Minimum	Median	Maximum	Average
All Industry - USA & Canada	32	.002	.32	7.50	1.05
Auto	0	-	-	-	-
Cement	0	-	-	-	-
Chemical	11	.02	.22	3.33	.75
Metal	1	.18	.18	.18	.18
Mining	0	-	-	-	-
Petroleum	5	.002	.07	.31	.12
Pulp and Paper	1	.33	.33	.33	.33
Rubber and Plastics	2	.28	.50	.71	.50
Textile	2	.07	1.00	1.92	1.00
Other Light Manufacturing	4	.09	1.10	2.80	1.27
Other Heavy Manufacturing	1	1.87	1.87	1.87	1.87
Other	5	.25	.94	7.50	2.86
Foreign	1	.33	.33	.33	.33

TABLE 25 - PLANT OUTAGE COST PER HR. DOWNTIME PER kW OF MAXIMUM DEMAND - PLANTS MORE THAN 1,000 kW MAX. DEMAND ($ per kWh)

Industry	Number of Plants Reporting	Minimum	Median	Maximum	Average
All Industry - USA & Canada	31	.0009	.36	5.77	.94
Auto	0	-	-	-	-
Cement	0	-	-	-	-
Chemical	12	.0009	.14	2.11	.33
Metal	1	.55	.55	.55	.55
Mining	0	-	-	-	-
Petroleum	2	.04	1.24	2.43	1.24
Pulp and Paper	1	.07	.07	.07	.07
Rubber and Plastics	3	.28	.36	1.33	.66
Textile	1	.24	.24	.24	.24
Other Light Manufacturing	4	.33	.54	1.20	.65
Other Heavy Manufacturing	1	.93	.93	.93	.93
Other	6	.75	2.50	5.77	2.69
Foreign	1	.07	.07	.07	.07

TABLE 26 - PLANT OUTAGE COST PER FAILURE PER kW OF MAXIMUM DEMAND - PLANTS LESS THAN 1,000 kW MAX. DEMAND ($ per kW)

Industry	Number of Plants Reporting	Minimum	Median	Maximum	Average
All Industry - USA & Canada	10	.50	3.68	10.00	4.59
Auto	0	-	-	-	-
Cement	0	-	-	-	-
Chemical	0	-	-	-	-
Metal	1	4.67	4.67	4.67	4.67
Mining	0	-	-	-	-
Petroleum	0	-	-	-	-
Pulp and Paper	0	-	-	-	-
Rubber and Plastics	0	-	-	-	-
Textile	0	-	-	-	-
Other Light Manufacturing	2	.50	1.11	1.72	1.11
Other Heavy Manufacturing	7	1.67	5.00	10.00	5.57
Other	0	-	-	-	-
Foreign	0	-	-	-	-

TABLE 27 - PLANT OUTAGE COST PER HR. DOWNTIME PER kW OF MAXIMUM DEMAND - PLANTS LESS THAN 1,000 kW MAX. DEMAND ($ per kWh)

Industry	Number of Plants Reporting	Minimum	Median	Maximum	Average
All Industry - USA & Canada	10	.86	4.42	27.00	8.11
Auto	0	-	-	-	-
Cement	0	-	-	-	-
Chemical	0	-	-	-	-
Metal	1	1.33	1.33	1.33	1.33
Mining	0	-	-	-	-
Petroleum	0	-	-	-	-
Pulp and Paper	0	-	-	-	-
Rubber and Plastics	0	-	-	-	-
Textile	0	-	-	-	-
Other Light Manufacturing	2	.86	1.43	2.00	1.43
Other Heavy Manufacturing	7	3.33	7.69	27.00	11.00
Other	0	-	-	-	-
Foreign	0	-	-	-	-

TABLE 28 - PLANT RESTART TIME (After Service is Restored Following a Failure that has Caused Complete Plant Shutdown)

Industry	Number of Plants Reporting	Average (Hours)	Median (Hours)
All Industry - USA & Canada..	43	17.4	4.0
Auto........................	0	-	-
Cement......................	0	-	-
Chemical....................	19	20.7	20
Metal.......................	1	4	4
Mining......................	0	-	-
Petroleum...................	3	37.3	24
Pulp and Paper..............	1	10	10
Rubber & Plastics...........	3	2.33	2
Textile.....................	3	58.3	72
Other Light Manufacturing....	7	2.14	2
Other Heavy Manufacturing....	1	2	2
Other.......................	5	2.6	1
Foreign.....................	1	48	48

TABLE 29 - CRITICAL SERVICE LOSS DURATION (Maximum Length of Power Failure that Will Not Stop Plant Production)

Industry	Number of Plants Reporting	Average	Median
All Industry - USA & Canada....	55	12.6 min.	10.0 sec.
Auto..........................	0	-	-
Cement........................	0	-	-
Chemical......................	20	4.56 min.	1.25 sec.
Metal.........................	2	15.0 min.	15.0 min.
Mining........................	0	-	-
Petroleum.....................	1	1.0 sec.	1.0 sec.
Pulp and Paper................	1	10.0 cycles	10.0 cycles
Rubber & Plastics.............	3	30.0 sec.	20.0 sec.
Textile.......................	3	3.34 min.	30.0 cycles
Other Light Manufacturing......	7	10.3 min.	10.0 sec.
Other Heavy Manufacturing......	10	47 min.	45 min.
Other.........................	8	1.9 min.	20.0 cycles
Foreign.......................	1	15.0 cycles	15.0 cycles

TABLE 30 - LOADS LOST VERSUS TIME OF POWER OUTAGE
(Tabulation of the Percentage of Equipment Failures for Which the Designated Load was Lost and Average Plant Outage Duration Resulting from these Failures)

	For Equipment Failures 1 Cycle or less in Duration			For Equipment Failures Between 1 and 10 Cycles in Duration			For Equipment Failures 10 Cycles or More in Duration		
Type of Load	Yes	No	Not Known	Yes	No	Not Known	Yes	No	Not Known
Computer	0%	0%	0%	4%	96%	0%	9%	91%	0%
Motor	0%	0%	0%	33%	67%	0%	67%	33%	0%
Lighting	0%	0%	0%	22%	78%	0%	38%	61%	2%
Solenoid	0%	0%	0%	22%	74%	4%	25%	66%	9%
Other	0%	0%	0%	7%	15%	78%	25%	62%	13%
Average Plant Outage Duration	0.0 Hours			1.39 Hours			22.6 Hours		

Only non-zero data was used in computing the average plant outage duration

5) Additional information on the cost of power outages in Sweden, Norway, and the United States is contained in [2].

Plant Restart Time (*Table 28*)

The textile, petroleum, and chemical industries have a much longer plant restart time than the other industries included in the survey.

Critical Service Loss Duration (*Table 29*)

1) There is a wide spread in critical service loss duration time for the 55 plants in the survey.

2) It is suggested that the data from the 55 plants should be compiled to show several percentiles (10, 25, 75, and 90 percent) in addition to the median value already tabulated.

Loads Lost Versus Time of Power Outage (*Table 30*)

1) An outage between 1 to 10 cycles resulted in 33 percent of the plants losing motor loads and 22 percent losing a solenoid and only 4 percent losing a computer load. An outage greater than 10 cycles resulted in 67 percent of the plants losing motor loads and 25 percent losing a solenoid and only 9 percent losing a computer load; many plants must not have had computer loads to give such a low value. In fact, many plants must not have had motor loads or solenoid loads either. The important parameter to look at is the change in these percentages from 0 to the maximum value as the length of power outage time is increased.

2) It is suggested that loss of load data be compiled for the following additional categories of outage duration time:

a) 10 to 15 cycles,
b) 15+ to 30 cycles,
c) 0.5 + to 2.0 s,
d) 2.0+ to 4.0 s,
e) greater than 4.0 s.

The average plant outage duration should also be determined for these categories.

REFERENCES

[1] IEEE Committee Report, "Report on reliability survey of industrial plants; Part I: Reliability of electrical equipment," this issue, pp. 213-235.

[2] R. B. Shipley, A. D. Patton, and J. S. Denison "Power reliability cost vs worth," *IEEE Trans. Power App. Syst.*, vol. PAS-91, pp. 2204-2212, Sept./Oct. 1972.

Report on Reliability Survey of Industrial Plants, Part III: Causes and Types of Failures of Electrical Equipment, the Methods of Repair, and the Urgency of Repair

IEEE COMMITTEE REPORT

Abstract—**An IEEE sponsored reliability survey of industrial plants was completed during 1972. This included the causes and types of failures of electrical equipment, the methods of repair, and the urgency of repair. The results are reported from the survey of 30 companies covering 68 plants in nine industries in the United States and Canada. This information is useful in the design of industrial power distribution systems.**

Introduction

A KNOWLEDGE of the causes and types of failures of electrical equipment is useful in the design of industrial power distribution systems. In addition it is also useful to know the failure repair method, whether or not the repair was urgent, and how long it had been since the previous maintenance had been performed. During 1972 the Reliability Subcommittee of the IEEE Industrial and Commercial Power Systems Committee completed a reliability survey of industrial plants. This is the third paper reporting results from the survey. Included in this paper are the results for 14 main classes of electrical equipment on

1) failure repair method;
2) failure repair urgency;
3) failure, months since maintained;
4) failure, damaged part;
5) failure type;
6) suspected failure responsibility;
7) failure initiating cause;
8) failure contributing cause;
9) failure characteristic.

The failure repair method includes either the repair of the failed component or the replacement of the failed component with a spare. This can have a significant effect on the average downtime per failure, and thus is an important factor in reliability and availability calculations.

Paper TOD-73-158, approved by the Industrial and Commercial Power Systems Committee of the IEEE Industry Applications Society for presentation at the 1973 Industrial and Commercial Power Systems Technical Conference, Atlanta, Ga., May 13-16. Manuscript released for publication November 5, 1973.

Members of the Reliability Subcommittee of the IEEE Industrial and Commercial Power Systems Committee are W. H. Dickinson, *Chairman*, P. E. Gannon, M. D. Harris, C. R. Heising, D. W. McWilliams, R. W. Parisian, A. D. Patton, and W. J. Pearce.

The failure repair urgency also has a significant effect on the average downtime per failure and thus is an important factor in reliability and availability calculations.

A preventive maintenance program can have an effect on the failure rate of electrical equipment. Thus a knowledge of whether or not maintenance has been performed recently prior to the failure is a significant factor in helping to determine whether or not the maintenance program is adequate.

The damaged part from a failure is of interest. In addition, a knowledge is also desirable of the type of failure, initiating cause, contributing cause, and suspected responsibility. This information is useful for correcting deficiencies in electrical equipment and electrical systems.

The failure characteristic can be defined as the effect that the failure has on the electrical system. Thus this information is very important.

Survey Form

The survey form used is shown in Appendix A of Part 1 [1]. All of the information reported on in this paper came from card type 3, columns 30-46. The definitions of *failure* and *repair time* are given in Part 1 [1].

Response to Survey

A total of 30 companies responded to the survey questionnaire, reporting data on 68 plants from nine industries in the United States and Canada. Every failure report on card type 3 did not have filled in all the information called for in columns 30-46. Tables 31 and 32 give the data for each main equipment class on how many failures had the information called for in columns 30-46. Each main equipment class contains 18 or more failures; this is believed to be an adequate statistical sample size.

Statistical Analysis

The results were compiled for 14 main equipment classes. The number of failures were tabulated for each category of each column (30-46, card type 3). This was then divided by the total failures in each column so as to give the percentage for each category for each column (for each main equipment class).

Survey Results

The results are tabulated for the 14 main equipment classes in Tables 33-41. Each table represents one column (of 30-46, card type 3).

Summary of Conclusions

Transformers

In the cases reported, there were approximately an equal number of incidences of repairing the failed transformer and replacing it with a spare. The repair urgency slightly favored a round-the-clock repair over the regular work-day schedule. Inadequate preventive maintenance did not seem to have much influence on the reported failures since no preventive maintenance was reported on only 5 percent of the failures; 11 percent of the failures were blamed on inadequate maintenance. Damaged insulation both in the windings and bushings accounted for the majority of the transformer damage, with the majority of failures being flashovers involving ground. 24 percent of the reported cases considered normal deterioration from age as the contributing cause of the failure, yet 39 percent reported that they felt the manufacturer was primarily responsible. Transient overvoltages, from lightning or switching surges, and other insulation breakdown account for 41 percent of the reported failures. In 90 percent of the reported cases the transformers were removed from the system by automatic protective devices; only 7 percent had manual removal.

Circuit Breakers

About the same number of circuit breakers were repaired in place as were replaced by spares. The relative importance of circuit breakers was indicated by 73 percent of the survey respondents making repairs on a round-the-clock basis. The bulk of the reported failures involved flashovers to ground with damage primarily to the protective device components and the device insulation. Transient overvoltages, insulation breakdowns, and protective device malfunctions were considered a major initiating cause with normal deterioration from age and misoperation or testing errors considered as contributing causes. However, 33 percent of the respondents could not classify the initiating cause into any of the survey classes, and 55 percent could not classify the contributing cause into any of the survey classes. In addition, 36 percent of the suspected causes of failure were blamed on "other." 42 percent of the reported failures involved circuit breakers opening when they should not; it is possible that several of these failures were external to the circuit breaker and of unknown cause and were blamed on the circuit breaker. 32 percent of the reported failures involved circuit breakers that failed during a load-carrying condition.

23 percent of the failures were blamed on the manufacturer and another 23 percent on inadequate maintenance, but 36 percent were blamed on "other." Inadequate preventive maintenance (PM) could be a factor of some significance since no PM was reported on 16 percent of the failures.

Motor Starters

Of the reported motor starter failures, about two thirds were repaired by replacing the starter with a spare and two thirds were repaired on a round-the-clock basis. About half of the cases reported indicate that the damage was other than the classes listed in the survey, primarily resulting from flashovers or electrical defects. 64 percent felt that a malfunction of a protective relay control device initiated the failure with 40 percent of the respondents reporting that normal deterioration from age was a contributing cause. Over half of the respondents felt that improper application was primarily responsible for the failure. In the cases reported 36 percent had been discovered during testing or maintenance, and 20 percent were only partial failures. Lack of preventive maintenance was not a big problem. Those starters that had been maintained less than 12 months prior to the failure accounted for 67 percent of the cases reported.

Motors

Of the reported motor failures, about three quarters were repaired versus about one fourth being replaced by a spare. About three quarters were repaired on a regular work-day basis. The types of failures varied from flashovers to electrical defects, to mechanical defects, with winding insulation and bearings sustaining the majority of the damage. Insulation breakdown, overheating, and mechanical seizing were blamed as the primary initiating causes with normal deterioration from age, loss or deficiency of lubricant, exposure to abnormal moisture, and exposure to aggressive chemicals ranking high on the list of contributing causes. 30 percent of the failures were discovered during testing or maintenance, which probably resulted in less actual damage in those cases. Inadequate maintenance, improper application, and defective equipment were listed as having primary responsibility. However, over half of the respondents could not assign responsibility into one of the survey classes. The motors that had been maintained between 12 and 24 months prior to the failure accounted for 57 percent of the reported cases with less than 12 months and more than 24 months accounting for 22 percent and 19 percent, respectively. No preventive maintenance accounted for only 2 percent, yet this does not correlate well with inadequate maintenance being listed as having primary responsibility in 17 percent of the reported cases.

Generators

Of the reported generator failures 84 percent were repaired in place. About the same number were repaired on a round-the-clock basis as were repaired on a regular work-day basis. 69 percent of the respondents reported damage other than the survey classes with electrical auxiliaries, winding insulation, and moving parts sustaining some damage. Mechanical breaking, transient overvoltages, and about half unclassified items were considered the primary initiating causes with normal deterioration from age and persistent overloading considered contributing causes. Responsibility was spread between inadequate maintenance and defective components with about half of the respondents unable to place primary responsibility into any of the survey classes. Infrequent or no preventive maintenance were not involved in any of the reported cases, a point that does not correlate with the fact that some of the respondents felt inadequate maintenance was the primary responsibility.

Disconnect Switches

Of the reported disconnect switch failures, 70 percent were repaired by replacement with a spare, with work in 80 percent of the cases being performed on a regular work-day schedule. Electrical defects, mechanical defects, and flashovers to ground resulted in damage to mechanical components and insulation. Some form of mechanical breaking or contact from foreign

TABLE 31 - NUMBER OF FAILURES FOR ELECTRIC UTILITY POWER SUPPLIES THAT CONTAINED THE INFORMATION CALLED FOR IN COLUMNS 30-46, CARD - TYPE 3

Card Type 3 Column	Title	Number of Failures
30	Failure Repair Method............	28
32	Failure Repair Urgency...........	35
34	Failure, Months Since Maintained..	25
36	Failure, Damaged Part............	39
38	Failure Type.....................	49
40	Suspected Failure Responsibility..	43
42	Failure Initiating Cause.........	53
44	Failure Contributing Cause.......	53
46	Failure Characteristic...........	145

TABLE 32 - NUMBER OF FAILURES FOR EACH MAIN EQUIPMENT CLASS THAT CONTAINED THE INFORMATION CALLED FOR IN COLUMNS 30-46, CARD-TYPE 3

Main Equipment Class	Maximum	Minimum	Avg.
Transformers	101	97	100
Circuit Breakers	176	161	171
Motor Starters	88	88	88
Motors	561(col.36)	493(col.40)	517
Generators	83(col.36)	31(all other)	37
Disconnect Switches	101	100	101
Swgr. Bus-Insulated	20	20	20
Swgr. Bus-Bare	24	20	23
Bus Duct	20	18	20
Open Wire	109	104	108
Cable	223	211	218
Cable Joints	45	44	45
Cable Terminations	51	47	50

sources accounted for about half of the initiating causes, with exposure to dust and contaminants and a large number of unclassified items considered contributing causes. Inadequate operating procedures, inadequate maintenance, and defective components were considered primarily responsible, which seems to correlate with over 66 percent of the reported cases not having any preventive maintenance and 21 percent not having any preventive maintenance 24 months prior to the failure.

Switchgear Bus, Bare

Of the reported uninsulated switchgear bus failures, about two thirds were repaired in place, with a little more than half of them being repaired on a round-the-clock basis. 79 percent of the respondents report some form of insulation damage all resulting from flashovers either to ground (79 percent) or between phases (21 percent). Mechanical failure, shorting by metal objects, and insulation breakdown were the predominant initiating causes with exposure to abnormal moisture, exposure to dust, exposure to aggressive chemicals, and normal deterioration due to age listed as contributing causes. Interestingly, 15 percent of the respondents listed misoperation or testing errors as a contributing cause. 39 percent felt that an outside agency was responsible for the failure, while 22 percent blamed inadequate maintenance.

Switchgear Bus, Insulated

Of the reported insulated switchgear bus failures, essentially all were repaired in place with over two thirds of the repairs being completed on a round-the-clock basis. 90 percent of the respondents reported insulation damage resulting primarily from flashovers to ground and between phases. Insulation breakdown was considered to have initiated the failure in about half of the cases, with exposure to contaminants, moisture, severe weather, and normal deterioration from age being considered as contributing factors. Improper application (45 percent) and inadequate maintenance (35 percent) were held responsible for the failures.

Bus Duct

Of the reported bus duct failures, 65 percent were repaired in place with the majority of them being repaired on a round-the-clock basis. 90 percent of the respondents reported some form of damaged insulation resulting from a flashover to ground. Mechanical failure, insulation breakdown, and overheating were blamed as initiating factors, with normal deterioration due to age being listed as a contributing factor in half of the cases. Responsibility for the reported failures varied from defective components (26 percent), improper application (16 percent), to inadequate maintenance (16 percent).

Open Wire

Of the reported open-wire failures, 70 percent were repaired in place with a little over half involving a round the clock effort. About half of the failures involved flashovers either to ground or between phases and about 25 percent involved other electrical defects. In the reported failures, transient overvoltages, overheating, or shorting by metal objects were considered the most significant initiating causes, with severe weather and exposure to aggressive chemicals being the predominant contributing causes. 81 percent of the respondents indicated that no preventive maintenance had been performed in over two years, which supports the fact that over a third of them blamed inadequate maintenance as being responsible.

Cables

The relative importance of primary cable was again indicated by about two thirds of the reported cases making repairs on a round-the-clock basis. There were a few more reported cases where repairs to cables were made by complete replacement rather than by in-place repairs. About three quarters of the failures involved flashovers to ground, resulting in insulation damage.

TABLE 33 - FAILURE REPAIR METHOD
TABLE 34 - FAILURE REPAIR URGENCY

ELECTRIC UTILITY POWER SUPPLIES	TRANSFORMERS	CIRCUIT BREAKERS	MOTOR STARTERS	MOTORS	GENERATORS	DISCONNECT SWITCHES	SWITCHGEAR BUS-INSULATED	SWITCHGEAR BUS-BARE	BUS DUCT	OPEN WIRE	CABLE	CABLE JOINTS	CABLE TERMINATIONS	Table, Title, Category
%	%	%	%	%	%	%	%	%	%	%	%	%	%	TABLE 33 - FAILURE REPAIR METHOD (Col. 30)
50	47	51	33	78	84	30	95	71	65	70	47	87	60	1. Repair of failed component in place or sent out for repair
46	53	49	67	22	16	70	5	29	35	9	53	13	34	2. Repair by replacement of failed component with spare
4	0	0	0	0	0	0	0	0	0	21	0	0	6	99. Other
														TABLE 34 - FAILURE REPAIR URGENCY (Col. 32)
91	51	73	66	23	48	20	70	58	80	55	66	56	53	1. Requiring round-the-clock all out efforts
9	45	22	34	74	52	80	25	33	15	26	28	22	31	2. Requiring repair work only during regular workday, perhaps with some overtime
0	4	5	0	2	0	0	5	8	5	0	6	22	16	3. Requiring repair work on a non-priority basis
0	0	0	0	0	0	0	0	0	0	19	0	0	0	99. Other

TABLE 35 - FAILURE, MONTHS SINCE MAINTAINED
TABLE 36 - FAILURE, DAMAGED PART

Table, Title, Category	ELECTRIC UTILITY POWER SUPPLIES	TRANSFORMERS	CIRCUIT BREAKERS	MOTOR STARTERS	MOTORS	GENERATORS	DISCONNECT SWITCHES	SWITCHGEAR BUS-INSULATED	SWITCHGEAR BUS-BARE	BUS DUCT	OPEN WIRE	CABLE	CABLE JOINTS	CABLE TERMINATIONS
TABLE 35 - FAILURE, MONTHS SINCE MAINTAINED (Col. 34)														
1. Less than 12 months ago	56	34	18	67	22	58	8	10	35	25	1	11	18	12
2. 12-24 months ago	40	38	60	17	57	42	5	35	30	45	8	13	20	12
3. Over 24 months ago	4	22	5	16	19	0	21	55	13	10	81	10	2	36
4. No preventive maintenance	0	5	16	0	2	0	66	0	22	20	9	66	60	40
99. Other	0	0	0	0	0	0	0	0	0	0	0	0	0	0
TABLE 36 - FAILURE, DAMAGED PART (Col. 36)														
1. Insulation - winding	0	68	0	5	50	7	0	0	0	15	0	5	0	0
2. Insulation - bushing	8	13	2	0	0	0	1	5	8	10	1	0	0	12
3. Insulation - other	10	3	19	10	3	0	14	90	71	65	6	84	91	75
4. Mechanical - bearings	0	0	1	0	29	2	0	0	0	0	0	3	0	0
5. Mechanical - other moving parts	3	0	11	16	3	7	9	0	0	0	0	0	0	0
6. Mechanical - other	15	1	6	2	1	4	30	0	0	0	4	1	0	4
7. Other electrical - auxiliary device	10	3	6	13	3	10	8	5	0	0	3	1	0	0
8. Other electrical - protective device	10	1	28	2	0	1	1	0	0	0	3	1	0	0
9. Tap changer - no load type	0	7	1	0	0	0	0	0	0	0	0	0	0	0
10. Tap changer - load type	0	1	0	0	0	0	0	0	0	0	0	0	0	0
99. Other	44	3	26	52	11	69	38	0	21	10	84	6	9	10

TABLE 37 - FAILURE TYPE

ELECTRIC UTILITY POWER SUPPLIES	TRANSFORMERS	CIRCUIT BREAKERS	MOTOR STARTERS	MOTORS	GENERATORS	DISCONNECT SWITCHES	SWITCHGEAR BUS-INSULATED	SWITCHGEAR BUS-BARE	BUS DUCT	OPEN WIRE	CABLE	CABLE JOINTS	CABLE TERMINATIONS	Table, Title, Category
														TABLE 37 - FAILURE TYPE (col. 38)
43	58	33	14	28	19	15	65	79	70	34	73	70	55	1. Flashover or arcing involving ground
4	13	10	20	4	3	4	35	21	30	23	1	9	4	2. All other flashover or arcing
14	12	19	55	32	29	47	0	0	0	25	7	20	37	3. Other electrical defect
8	10	11	11	31	32	14	0	0	0	6	5	0	4	4. Mechanical defect
31	7	27	0	6	16	21	0	0	0	12	14	0	0	99. Other

TABLE 38 - SUSPECTED FAILURE RESPONSIBILITY

ELECTRIC UTILITY POWER SUPPLIES	TRANSFORMERS	CIRCUIT BREAKERS	MOTOR STARTERS	MOTORS	GENERATORS	DISCONNECT SWITCHES	SWITCHGEAR BUS - INSULATED	SWITCHGEAR BUS - BARE	BUS DUCT	OPEN WIRE	CABLE	CABLE JOINTS	CABLE TERMINATIONS	Table, Title, Category
%	%	%	%	%	%	%	%	%	%	%	%	%	%	TABLE 38 - SUSPECTED FAILURE RESPONSIBILITY
8	39	23	18	15	19	29	5	9	26	0	16	0	0	1. Manufacturer-defective Component
0	0	0	0	0	0	0	0	0	0	0	0	0	0	2. Transportation to Site - defective handling
0	2	4	51	9	0	6	45	4	16	2	8	0	18	3. Application Engineering - improper application
0	3	3	0	1	3	4	10	17	5	9	14	50	39	4. Inadequate installation and testing prior to start-up
0	11	23	8	17	19	13	35	22	16	30	10	18	22	5. Inadequate maintenance
6	9	6	3	4	3	40	0	0	0	2	3	0	0	6 Inadequate operating procedures
17	2	5	0	0	0	1	0	22	5	5	4	5	0	7. Outside agency - personnel
32	4	1	0	1	6	0	0	17	0	21	6	2	8	8. Outside agency - other
38	30	36	19	53	48	8	5	9	32	31	38	25	14	99. Other

TABLE 39 - FAILURE INITIATING CAUSE

ELECTRIC UTILITY POWER SUPPLIES	TRANSFORMERS	CIRCUIT BREAKERS	MOTOR STARTERS	MOTORS	GENERATORS	DISCONNECT SWITCHES	SWITCHGEAR BUS - INSULATED	SWITCHGEAR BUS - BARE	BUS DUCT	OPEN WIRE	CABLE	CABLE JOINTS	CABLE TERMINATIONS	Table, Title, Category
														TABLE 39 - FAILURE INITIATING CAUSE(Col. 42)
33	23	13	1	6	10	4	5	5	0	26	26	11	12	1. Transient overvoltage disturbance (lightning, switching surges, arcing ground fault in ungrounded system)
0	0	0	0	0	0	0	0	0	0	0	0	0	0	2. Overvoltage
0	11	3	1	26	3	4	0	5	30	21	1	0	2	3. Overheating
5	18	18	8	30	3	5	50	18	20	8	29	40	51	4. Other insulation breakdown
7	17	13	8	4	29	17	10	23	45	7	24	31	24	21. Mechanical breaking, cracking, loosening, abrading or deforming of static or structural parts
2	0	5	6	20	3	2	0	0	0	0	0	0	0	22. Mechanical burnout, friction, or seizing of moving parts.
14	1	1	0	3	3	20	0	0	0	10	7	0	4	23. Mechanically caused damage from foreign source (digging, vehicular, accident,etc.)
12	1	2	5	0	0	0	0	23	5	14	2	0	2	41. Shorting by tools or metal objects
2	2	1	1	0	0	0	0	9	0	3	0	0	2	42. Shorting by birds, snakes, rodents, etc.
0	0	1	0	0	3	0	0	0	0	0	0	0	0	51. Loss of control power
2	1	11	64	5	0	0	0	0	0	0	0	0	0	52. Malfunction of protective relay control device, or auxiliary device.
0	0	0	0	0	0	3	0	0	0	0	0	0	0	61. Low voltage
2	0	0	0	0	0	0	0	0	0	0	0	0	0	62. Low frequency
21	25	33	7	5	45	45	35	18	0	11	10	18	4	99. Other

TABLE 40 - FAILURE CONTRIBUTING CAUSE

ELECTRIC UTILITY POWER SUPPLIES	TRANSFORMERS	CIRCUIT BREAKERS	MOTOR STARTERS	MOTORS	GENERATORS	DISCONNECT SWITCHES	SWITCHGEAR BUS - INSULATED	SWITCHGEAR BUS - BARE	BUS DUCT	OPEN WIRE	CABLE	CABLE JOINTS	CABLE TERMINATIONS	Table, Title, Category
														TABLE 40 - FAILURE CONTRIBUTING CAUSE (Col. 44)
2	13	4	0	5	10	8	0	0	6	0	2	0	0	1. Persistent overloading
4	0	1	0	1	6	3	5	0	0	0	0	2	0	2. Above-normal temperatures
0	0	0	0	0	0	1	0	0	0	0	0	0	0	3. Below-normal temperature
0	0	2	0	7	0	0	0	10	0	28	14	13	10	4. Exposure to aggressive chemicals or solvents
2	6	3	0	10	6	4	15	20	17	1	8	22	12	5. Exposure to abnormal moisture or water
0	0	0	0	0	3	0	0	5	0	3	2	0	0	6. Exposure to non-electrical fire or burning
0	0	0	0	2	0	0	0	0	0	0	1	0	0	8. Obstruction of ventilation by foreign objects or material
4	24	17	40	34	32	5	20	10	50	3	30	29	24	9. Normal deterioration from age
38	6	1	0	2	3	0	20	5	11	30	15	2	16	10. Severe wind, rain, snow, sleet, or other weather conditions
2	0	2	0	0	6	0	0	0	0	1	0	0	0	11. Protective relay improperly set
0	0	1	2	15	0	0	0	0	0	0	0	0	0	12. Loss or deficiency of lubricant
0	0	0	0	1	0	0	0	0	0	0	0	0	0	13. Loss or deficiency of oil or cooling medium
0	3	10	3	0	0	0	0	15	6	2	3	0	8	14. Misoperation or testing error
4	3	3	1	5	0	26	40	20	0	2	1	0	0	15. Exposure to dust or other contaminants
45	44	55	53	18	32	53	0	15	11	31	24	31	29	99. Other

TABLE 41 - FAILURE CHARACTERISTIC

ELECTRIC UTILITY POWER SUPPLIES	TRANSFORMERS	CIRCUIT BREAKERS	MOTOR STARTERS	MOTORS	GENERATORS	DISCONNECT SWITCHES	SWITCHGEAR BUS - INSULATED	SWITCHGEAR BUS - BARE	BUS DUCT	OPEN WIRE	CABLE	CABLE JOINTS	CABLE TERMINATIONS	Table, Title, Category
%	%	%	%	%	%	%	%	%	%	%	%	%	%	TABLE 41 - FAILURE CHARACTERISTIC (Col. 46)
														Utility Power Supplies (Select code)
10	1	0	0	0	0	0	30	8	10	0	17	0	12	1. Failure of single circuit (no redundant supply)
71	0	1	0	0	0	0	5	0	0	0	7	0	2	2. Failure of one circuit of a double-circuit redundant supply
15	0	0	0	0	0	0	0	0	0	0	0	0	0	3. Failure of both circuits of a double-circuit redundant supply
2	0	1	0	0	0	0	0	8	0	0	0	0	0	4. Failure of all circuits of a three or more circuit redundant supply
0	0	0	0	0	3	0	0	0	0	0	0	0	0	5. Partial failure of a three or more circuit redundant supply
														Transformers (Select Code)
0	90	0	0	0	0	4	0	0	0	0	4	0	2	6. AUtomatic removal by protective equipment
0	1	0	0	0	0	0	0	0	0	0	0	0	0	7. Partial failure reducing capacity
0	7	0	0	0	0	0	0	0	0	0	0	0	0	8. Manual removal

TABLE 41 - FAILURE CHARACTERISTIC

Table, Title, Category	ELECTRIC UTILITY POWER SUPPLIES	TRANSFORMERS	CIRCUIT BREAKERS	MOTOR STARTERS	MOTORS	GENERATORS	DISCONNECT SWITCHES	SWITCHGEAR BUS - INSULATED	SWITCHGEAR BUS - BARE	BUS DUCT	OPEN WIRE	CABLE	CABLE JOINTS	CABLE TERMINATIONS
Circuit Breakers (Select Code)														
9. Failed to close when it should	0	0	5	1	0	0	0	0	0	0	0	0	0	0
10. Failed while opening	1	0	9	1	0	0	0	0	0	15	0	0	0	0
11. Opened when it shouldn't	0	0	42	1	0	6	0	0	0	0	0	0	0	0
12. Damaged while successfully opening	0	0	7	0	0	0	0	0	0	0	0	0	0	0
13. Damaged while closing	0	0	2	0	0	0	0	0	0	0	0	0	0	0
14. Failed while operating (not while opening or closing)	0	0	32	0	0	0	0	0	0	0	0	0	0	0
General (Select Code for any other class)														
15. Failed (this applies to all classes	0	1	0	34	68	65	68	65	71	65	69	65	96	65
16. Failed during testing or maintenance	0	0	1	5	1	0	3	0	13	5	2	2	4	2
17. Damage discovered during testing or maintenance	0	0	1	36	30	0	18	0	0	0	1	2	0	12
20. Partial failure	0	0	0	20	0	16	6	0	0	5	6	3	0	6
99. Other	0	0	0	1	0	10	1	0	0	0	23	1	0	0

An interesting point is that in over two thirds of the failures there had been no preventive maintenance, yet inadequate maintenance was only listed in 10 percent of the cases as being responsible for the failure. 16 percent placed the responsibility with the manufacturer, 14 percent with inadequate installation and testing prior to start-up, with 38 percent of the cases reporting reasons for the failure in classes other than those listed in the survey.

The initiating causes varied from transient overvoltage disturbances to insulation breakdown, to mechanical failures, with 30 percent reporting normal deterioration from age as a contributing cause.

Cable Joints

Of the failures reported, 87 percent were repaired in place, with just over half being repaired on a round-the-clock basis. Almost all of the failures resulted in damaged insulation, primarily from flashovers to ground, which were initiated by insulation breakdowns, transient overvoltages, or mechanical failure.

29 percent of the respondents felt that normal deterioration from old age contributed to the failure, while 35 percent blamed abnormal moisture or exposure to aggressive chemicals. Inadequate installation and testing were considered responsible for 50 percent of the failures. 60 percent of the respondents reported that no preventive maintenance had been performed, but only 18 percent blamed the failure on inadequate maintenance.

Cable Terminations

Of the reported cable termination failures, 60 percent were repaired in place with just over half of the repairs being made on a round-the-clock basis. The primary damage was insulation involving either a flashover to ground or other electrical defect. About half of the respondents felt that the failure was initiated by an insulation breakdown, with normal deterioration due to age, severe weather, and exposure to abnormal moisture or aggressive chemicals contributing significantly to the problem. 39 percent felt that inadequate installation and testing prior to start-up was primarily responsible, while 22 percent felt that inadequate maintenance should be blamed. This also seems to correspond to the reporting that in 40 percent of the cases no preventive maintenance had been performed in over two years.

General Conclusions

Electrical Equipment

The general picture from Tables 38 and 35 spotlights inadequate maintenance as a significant factor in the suspected responsibility for failures. Yet the owner appears willing to work round the clock to fix failures after they have occurred. Lack of cleaning and lubrication is apparent on disconnect switches, buses, open wire, cable, cable joints, cable terminations, and motors.

Electric Utility Power Supplies

Many of the results shown in Tables 33–38 are not really applicable for electric utility power supplies because the questions asked are not well suited. The importance of the utility supply was indicated by 91 percent of respondents making repairs on a round-the-clock basis. The failures were predominantly flashovers involving ground, caused by lightning during severe weather or by dig-ins or vehicular accident. Outside agencies, probably the local utility, were predominantly responsible for the failure with preventive maintenance having no apparent effect on the cases reported.

The data reported under "failure characteristic" in Table 41 are of special significance in the case of double- or triple-circuit electric utility power supplies. In particular, the failure rate can

TABLE 42 - SIMULTANEOUS FAILURE OF ALL CIRCUITS IN ELECTRIC UTILITY POWER SUPPLIES

% of 145 Failures from Table 41	Number of Failures	Utility Power Supplies - Failure Characteristic from Table 41
15%	22	3. Failure of both circuits of a double-circuit redundant supply
2%	3	4. Failure of all circuits of a three or more circuit redundant supply
17%	25	Total number of simultaneous failures of all circuits in a double or more circuit redundant supply

be calculated for the simultaneous failure of all circuits in a double- or triple-circuit electric utility power supply.

From Table 3 of Part 1 [1] the sample size is 210.7 unit-years for a double- or triple-circuit electric utility power supply. A double- or triple-circuit supply operating for one year is counted as one unit-year. It is possible to calculate a failure rate from these data as follows:

$$\frac{25}{210.7} = 0.119$$ failures per year for simultaneous failure of all circuits in a double- or triple-circuit electric utility power supply.

Some discrepancies were found in the data on the number of installed units for double- and triple-circuit electric utility power supplies. See the discussion in Part 1 [1] on this point.

Discrepancies

A survey such as this one often obtains some data that appear to contain errors. Sometimes the results look ridiculous. However, some of the ridiculous looking results may actually be correct. Some of the errors are believed due to a misinterpretation of the question by the respondent.

The data in Tables 31–41 have been published without attempting to correct discrepancies or errors. A brief list of some possible discrepancies is given.

Table 36: The damaged part of one percent of failed circuit breakers is a tap changer. The damaged part of three percent of failed cables is a bearing. Winding insulation is shown as the damaged part in failures of cables, bus ducts, and motor starters.

Table 39: Three percent of the failures in disconnect switches were initiated by low voltage.

REFERENCES

[1] IEEE Committee Report, "Report on reliability survey of industrial plants, Part I: Reliability of electrical equipment," this issue, pp. 213–235.

Discussion

J. Krasnodebski, N. M. Thompson, D. H. Cooke, A. W. W. Cameron, S. Basu, and T. J. Ravishanker (Ontario Hydro, Toronto, Ont., Canada):

1) *Quality of Input Data:* The confidence level of data in a survey of this kind cannot be assessed by mathematics only. One key problem is the adequacy of records and completeness of data. Some of the apparent discrepancies noted in the paper seem to indicate quite substantial omissions in records. Unless the industries involved keep much better failure records than we have done to date, this is not surprising. The first requirement of a useful reliability program is an adequately complete and accurate system for recording failures and consequences (in outage terms).

TABLE A
GENERATORS

Forced Outages

EEI Report

Sample Size (unit-years)	Number of Occurrences per Unit-Year	Outage Hours per Occurrence
204	0.142	91.8
404	0.839	126.5
705	0.521	54.4
483	0.393	125.6

IEEE Reliability Survey

Type of Drive	Sample Size (unit-years)	Number of Occurrences per Unit-Year	Outage Hours per Occurrence
Steam turbines*	761.8	0.032	165.0
Jet engines			
Gas turbines	89.4	0.638	23.1
Diesel engines	59.4	0.067	127.0

*EEI results are for generators 60–89 MW.

The requirements for better records, along with the detail involved in the report forms, indicate that acquiring useful data of this kind is time consuming.

It is suggested that, if a choice is necessary, it might be preferable to have a limited (but statistically adequate) number of plants establish a reliably complete recording and reporting system rather than increase the size of the sample under current record systems.

2) *Survey Results on Equipment Failures:* The failure rate is given in failures per unit-year. Is year in this context a calendar year or 8760 hours of plant or equipment operating time? If the failure rate is given per calendar year, were adjustments made for plants operating for 40 hours per week against those operating for up to 168 hours per week?

3) *Discussion of Equipment:*

Motors: It is suspected that the discrepancy in failure rates results from the different application of the two types of motors. Synchronous motors are usually applied only in engineered situations and are carefully designed for the application. Large synchronous motors are usually slow speed. Induction motors are mass produced, purchased off the shelf at the lowest cost, and usually operated to take advantage of any service factor. The survey figures are probably correct but cannot be used for comparison of reliability, leading to a conclusion that synchronous motors are more reliable. It is a comparison of apples and oranges.

Switchgear Bus: The paper states that the reported data are the opposite to what they should be. The reported figures may be correct. Manufacturers regularly reduce the spacing between buses and the spaces between phases and ground when they use insulated bus. As the conductor insulation is usually also reduced by design and occasionally by inferior material standards compared to that on insulated cables, and workmanship is frequently less than perfect, failures on this type of gear are probably at least as common as those on air-insulated equipment.

Circuit Breakers: The failure rate for circuit breakers appears much too low. It must of course be a function of the frequency of operation as well as lapsed time. We did not find a definition of circuit breaker failure, which we believe should differ from cable, transformer, or other static device failures. Circuit breaker failures should be based on failure to operate satisfactorily either to remain closed or to open or to close when called upon. It should be clear whether these figures include failures caused by auxiliaries such as instrument transformers, relays, and control switches. Since any calculation of the reliability of a power system would be made unreasonably complex by attempts to treat all these devices individually, a figure for circuit breaker failures which includes them is usually required by the designer.

Generators: For the generators in the electrical power industry a good source of data exists in the EEI "Report on Equipment Availability for Twelve-Year Period 1960–1971." The comparison between the failure rates and average repair time contained in that report and the survey discussed are shown in Table 43. EEI data quoted for steam turbine driven generators are for the size class 60–89 MW, which is probably larger than the average size of a corresponding generator included in the industrial survey.

It can be seen that the EEI failure rate for steam turbine driven generators based on forced outages is higher by a factor of 5 than in the industrial survey. For gas turbines, failure rates contained in both reports are of the same order, while the outage duration quoted in the EEI report is higher. 54 failures in 5.5 unit years in the petroleum industry can probably be explained by the start-up troubles.

In summary, experience in the utility industry seems to explain results obtained in the industrial survey to a large degree.

4) *Causes of Failure:*

a) How important is the age of equipment? It is mentioned only as a "contributing cause," second in frequency only to "other." Are there economic replacement times, or does obsolescence usually come first?

b) Should the inference be drawn that reliability of industrial equipment, which is reasonably well suited to its job, depends mainly on 1) stringent acceptance testing, especially overvoltage testing, 2) adequate cleaning, and 3) proper lubrication of bearings?

5) *Additional Suggestions for Analysis:* Consideration should be given to add the manufacturer of the main class of equipment to provide information on reliability of different manufacturers.

Carl Becker (Cleveland Electric Illuminating Company, Cleveland, Ohio 44101): The Reliability Subcommittee did an outstanding job in assembling and correlating the mountainous volume of data in a simple, easy to understand tabulation. I would like to add some discussion that I feel would help the value of these tables and add to the accuracy of future studies. My two main points are 1) the downtime per failure on a single-circuit utility supply is extremely high (possibly by a factor of five), and 2) the equation for the dollars lost per interruption may be improved by using other than the kilowatt demand and kilowatthour usage as bases.

My company gathers, codes, and analyzes by computer all interruptions to our three quarter million customers. The average downtime per customer on our distribution system (which is a single-circuit radial supply) has been between 51 and 61 min for five of the past six years. Our service area experienced a catastrophic storm during 1969 which caused the average downtime per customer to jump to 124 min. In addition, my company is of the opinion that no plant should be down for more than 4 h (barring major catastrophies). A report is therefore written for each interruption exceeding 4 h in duration, and these reports are extremely few in number. Furthermore, 13 utilities have polled their reliability statistics for customers fed from the distribution system and found the average downtime per interruption for 1971 to be approximately 1½ h long. The average downtimes ranged from 0.75 to 3.2 h.

This information shows that the downtime per failure for industrial plants is probably outside the predicted tolerance on the IEEE data. This variance may be due to either a major long disturbance affecting a majority of those industrial plants participating or to misinterpretation of the information required.

For over five years I have worked with our customers in regard to reliability problems. My experience has shown that the plant investment, labor cost, and value of product is a better gauge of the cost per minute down than would be either maximum kilowatthour demand or usage. For example, I worked with a manufacturer of magnesium parts for military aircraft (I will call this plant A) and another manufacturer of parts for conveyor systems (plant B). The dollar loss for A per minute down was 100 times greater than that for B. However, plant B's demand is 2500 kW and A's demand is 500 kW, which is an indication that the kilowatthour consumptions in these particular cases are not related at all to the economical loss due to a power interruption. In general I find that the cost of downtime is tied heavily to one of the following: 1) the number of employees, 2) the cost of the product in production (piecework), or 3) the dollar output per hour (high production). A combination of these three items would indicate that loss is tied to the dollars out of the plant per unit of time. Therefore I feel that future studies should relate downtime to dollars per minute of plant production, gross plant, etc.

J. W. Beard (Union Carbide Corporation, South Charleston, W. Va. 25303): The report format and the manner in which the information is presented is generally quite adequate. Appendix A (Part I) is somewhat difficult to read because of the reduced print, but I am not suggesting it be upgraded for this report. Because of the many and various pieces of data used for the report, it is understandable that the reader must spend a great deal of time in studying and analyzing the information in order to properly apply it. The "readily" understandable factor should perhaps be given more consideration in defining the criteria for future surveys.

It is my opinion that the most useful types of information presented are:

1) failure rate and failure rate confidence limits;
2) failure, damaged part;
3) failure type;
4) failure initiating cause;
5) failure contributing cause;
6) failure characteristics.

I believe it is a good assumption that the raw data submitted for many of the other types of information represented were of much lesser accuracy than for these. For example, most plants reporting data for information types such as plant outage cost, critical service loss duration, and loads lost versus time of power outage probably had to draw on someone's memory of each failure and then apply the "best estimate" principle. This factor alone raises the question as to whether these types of information can ever be constructed to have useful meaning. Except for near catastrophic failures, which result in heavy financial losses, it is doubtful that most plants will spend the money to document this type of data. Furthermore, in a practical sense, when configuring systems and applying electrical equipment, the reliability requirement must be carefully considered for each producing unit served inasmuch as there are many variables that enter into the calculation of downtime losses.

The following suggestions are offered for consideration in any future surveys.

1) Basically concentrate on failure rates and failure causes.

2) Simplify and reduce scope of the survey questionnaire forms (present forms tend to scare users from contributing).

3) Omit asking for types of information such as cost of outage, repair time, plant start-up time, etc.

4) Instruct users *not to* report failures of equipment where reasonable preventative maintenance is not performed.

5) Instruct users *not to* report failures of misapplied equipment.

6) Instruct users *not to* include equipment installed prior to January 1, 1968.

7) Instruct users to give "in-service" date (energized) of all equipment units, not just on the reported failures.

8) Define "failure" as "damage to equipment sufficiently severe to force an outage by either manual or automatic removal of voltage." (Keep in mind that failures caused by the conditions in 4) and 5) are not to be reported.)

Part I: There seemed to be a great deal of confusion by the respondents on the information desired for electric power supplies. Thus the published failure rates may be questionable. It is my opinion that the questionnaire form for this was too nondescript. Perhaps one way to clearly describe the power supplies on which information is desired would be to include on the form simple single-line diagrams of the more common types of utility services.

It is my opinion that the lack of response by many companies was due primarily to poor and/or nonexistent records. A major contributing cause may have been the massive amount of information asked for.

The Reliability Subcommittee's judgement that a minimum of 8 to 10 observed failures was required for "good" accuracy when estimating equipment failure rates seems reasonable.

The value chosen for the confidence interval (0.90) was a good choice. The inclusion of confidence limits curves (Fig. 1) adds measurably to the report.

I generally concur with the Subcommittee's discussion comments. Their discussion of some of the results presented in the tables reinforces my feeling that the survey was too broad in scope, and the information submitted by the plants too ambiguous for meaningful interpretation.

While the sample sizes would be made smaller, as a general rule I feel that equipment should be grouped by voltage class. For example, in Table 2 one grouping of cable terminations is for 601–15 000 V. In this instance it would be especially helpful to know the failure rate on 15-kV cable terminations alone.

Part II: As stated in my general comments, I feel that it is not practical to generate reasonably accurate information of these types.

The bases for the units used in cost calculations, dollars per kilowatt plus dollars per killowatthour, are somewhat confusing. Clarification of this would be helpful.

In the Subcommittee's discussion of the cost of power outages, item 2), I must disagree with their thought that electrochemical or heating processes tend to have low outage costs because heat not supplied now can be supplied later.

In the discussion of loads lost versus time of power outage the "time" factor is questionable. Most plants are not equipped to measure short-duration power outages (cycles or even seconds).

Part III: Many of the information types in this part are very important. Some, I feel, are not. I suggest that the questions on failure repair method; failure repair urgency; failure, months since maintenance; and suspected failure responsibility be omitted from future surveys. The remaining types of information may be refined using knowledge gained from this survey.

In the Subcommittee's Summary of Conclusions they report that transient overvoltages were a major cause of failure in equipment such as, for example, transformers and circuit breakers; but I got the impression that much of this was speculation on the part of those responding. The possibility of transient overvoltage should be considered in the investigation of most equipment failures, and IEEE could perform an important service to industry by developing a socalled "evaluation of possibility of transient overvoltage contribution to equipment failures" guide.

Stanley Wells (Union Carbide Corporation, Port Lavaca, Tex. 77979): The Reliability Subcommittee should be congratulated for performing such a comprehensive reliability survey of industrial plants and for providing a very thorough report.

I would like to limit my discussion to Part 3 and, in particular, the preventive maintenance effect on the failure rate. A preventive maintenance program can very definitely have a direct effect on the failure rate of electrical equipment. In the modern automated plant of today, production demands and losses associated with downtime influence maintenance schedules. Equipment is often allowed to remain in operation for periods that exceed desired preventive maintenance time schedules. It is interesting to note that the survey indicates that preventive maintenance can be performed, yet equipment failures occur within a time period which is less than 12 months since preventive maintenance was performed. Our first attempt at a preventive maintenance program met with the same results. The program was reviewed in depth and it was found that it was inadequate and that the preventive maintenance procedures and time schedules should be reviewed and correlated with our failure experience. As experience was gained, the equipment preventive maintenance program developed into a very useful tool to practically eliminate electrical equipment failure. We soon recognized that where preventive maintenance periods were over 24 months or where no preventive maintenance at all was performed, chances of failure were extremely high. This fact is born out in the results of this survey. Table 35, "Failure–Months Since Maintained," has been rearranged to show that a large reduction in failures may be possible if preventive maintenance periods are on a 12- to 18-month basis (Table B).

Let's define preventive maintenance. Preventive maintenance is a system of routine inspections designed to minimize or forestall future equipment operating problems or failures, and which may, depending upon equipment type, require equipment exercising or proof testing. From this definition, the four following items listed under Table 38, "Suspected Failure Responsibility," can be considered a definite part of a maintenance program:

1) manufacture, defective components (locate by inspection or test);
2) application engineering, improper application;
3) inadequate installation and testing prior to start-up (proof test);
4) inadequate maintenance.

It is interesting to note that the survey indicates that these four items are responsible for a very large percentage of failures. The total for each category is listed below.

	Percent
Transformers	55
Circuit breakers	53
Motor starters	77
Motors	42
Generators	41
Disconnect switches	52
Switchgear bus insulated	95
Switchgear bus uninsulated	52
Bus duct	63
Open wire	41
Cable	48
Cable joints	68
Cable terminations	79

To increase the electrical system reliability, each failure should be very carefully analyzed to determine the failure cause, and corrective action to prevent additional failures should be applied to all applicable equipment.

TABLE B
FAILURES

	Less than 12 Months Ago Preventive Maintenance	12 Months or More or No Preventive Maintenance
Transformers	34	65
Circuit breakers	18	81
Motor starters	67	33
Motors	22	78
Generators	58	42
Disconnect switch	8	92
Switchgear bus insulated	10	90
Switchgear bus uninsulated	35	65
Bus duct	25	75
Open wire	1	98
Cable	11	89
Cable joint	18	82
Cable terminations	12	88

R. E. Kuehn (IEEE Reliability Group): The reliability, maintainability, and downtime logistics in the power area is very important and should lend itself to cost analysis, which is the ultimate judge of the value of reliability and maintainability programs. A great deal of data have been analyzed with all the obvious advantages and disadvantages that are entailed in such a data base. Parts 1 and 2 present me with a severe problem as a reliability professional and manager. In both papers a large effort was spent indicating that the survey results do not agree with what the engineering judgment says the results should be; for example, the discussion of Part 1 on motors, generators, cable, and switchgear bus. My quandary is that if I accept your judgment in all logic, I must question the validity of all the data collected, not just for motors, generators, cable, and switchgear bus. A possible procedure would have been to test the hypothesis that a part of the data was significantly different enough from the total grouped data to justify its rejection as part of the group data.

I would like to recommend analysis of variance or multiple regression in analyzing the data. It would appear that a number of possible variables exist and their effects are suitable for quantization. These procedures are covered in [1]-[4].

REFERENCES

[1] R. G. Stokes and F. N. Sehle, "Some life-cycle cost estimates for electronic equipments," in *Proc. 1968 Annu. Symp. on Reliability*, pp. 169-183.
[2] B. L. Retterer, "State of art assessment of reliability and maintainability as applied to ship systems," in *Proc. 1969 Annu. Symp. on Reliability*, pp. 133-145.
[3] H. Dagen, "Multiple regression," in *Proc. 1972 Annu. Symp. on Reliability*," pp. 51-58.
[4] "Cost effectiveness evaluation procedures for shipboard electronic equipment," ARINC Research Publ. 509-01-2-564 and 541-01-1-766.

Tai C. Wong (American Electric Power Service Corporation, New York, N.Y. 10004): The members of the Reliability Subcommittee are to be commended for conducting and analyzing the results of a survey that covers so many elements in industrial power systems.

Perhaps the authors want to clarify why the chi-squared distribution was used in fitting the data and what kind of statistical testing technique was employed to ensure the adequacy of the distribution chosen. The authors did compare the results of the recent survey against those obtained in 1962. The readers should be warned that this is only an observation based on empirical data and that any inference of a trend in the equipment reliability may not be valid. The paper indicates that many of the reported data cover more than one year of operating experience. Because the first survey was conducted twelve years ago, it is felt that the number of years that the different equipments were in service should be published (or the data collected during the next survey if they are not yet available) so that the reader can have a better understanding of the data background when he has to draw further conclusions, beyond the tables presented.

The authors indicated that the purpose of this survey is to make possible the quantitative reliability comparisons between alternative designs of new systems and then use this information in cost-reliability tradeoff studies to determine which type of power distribution system to use. It appears that the authors focus on making the economic tradeoff comparisons based on the available system components at a given time. However, the authors pointed out that the product of failure rate times the average downtime per failure is almost the same in 1973 as in 1962. Perhaps the equipment manufacturers and the industries can establish more dialogues, leading to an answer to the following two questions.

1) Should the equipments have a lower failure rate, but when failing, take longer to repair? or

2) Should the equipments have a higher failure rate, but when failing, need shorter repair time?

In a few instances during the survey, the respondents misinterpreted either the question(s) and/or the definition of the terms, thus leading to unreliable or biased results. This is especially true in the area of preventive maintenance. I might suggest that during the next survey 1) the definition of all terms that are likely to cause confusion in the questionnaire be included, 2) a pilot survey be instituted and any necessary modifications be made to the questionnaire before a full-scale survey is launched, or 3) the survey form be sent out without requesting data, but instead requesting the respondent's interpretations of the questions and the terms used. Then the survey form may be redesigned and data requested.

I. O. Sunderman (Lincoln Electric System, Lincoln, Nebr.): The authors have presented an interesting cross section of costs involved with industrial electric equipment downtime as accumulated by the computer. The data are to be utilized by interested parties in the choice of a reliability design for industrial power distribution systems. The wide range of costs as split into the two parts over 1000 kW and under 1000 kW suggests consideration of other kW brackets at 500, 2500, 5000, 7500, 10 000 kW, etc. The sufficiency of data will dictate breaking points, as the author already questions the cost data below 1000 kW.

In Part 3 the authors have reviewed and presented in excellent tables the results of electric equipment outage reports and repair. It must have been disturbing to note the numerous "other than categories classified." Perhaps further reporting on the "other" category comments, if available, would bring additional results to light.

IEEE Reliability Subcommittee: The authors wish to thank those who presented discussions on these three papers. Some of the suggestions given can be considered for incorporation into future surveys and they can also be used in the analysis of the results.

Several discussers have raised the question about the effect of "in service date" or age on the reliability of electrical equipment. Population data were collected on the average age of equipment in service; these will be published in Part 4. However, the Reliability Subcommittee did not request these data in the survey questionnaire on equipment failures. This subject was considered by the Subcommittee when making up the questionnaire; it was not included because this would have added additional complications to a questionnaire that was already considered too long. This meant that the assumption was made that the failure rate was constant with age. Thus a chi-squared distribution is appropriate for use in calculating the confidence limits of the failure rate. The assumption of a constant failure rate with age can be justified for most electrical equipment based upon reliability surveys made by others.

Mr. Becker and Mr. Beard have raised questions about the accuracy of the cost of power outage data and the attempt to relate it to kilowatts and kilowatthours. Information was collected but not published on the estimated plant outage costs 1) per failure and 2) per hour of downtime. The authors consider that the cost of power outages is an important factor that should be considered in the design of power distribution systems for industrial plants. Since power distribution systems are designed on the basis of kilowatt capacity and kilowatthour of delivered energy, it was felt that it is necessary to attempt to relate the cost of power outages to these two parameters. The approach used by the Reliability Subcommittee is the same as that which has been used by electric power companies in several European countries. The survey result of the median cost of 83¢ per kilowatthour of undelivered energy is in the same range as values obtained from surveys that have been made in Sweden, Norway, France, Italy, and West Germany. The authors agree that the published data of the cost of power outages are more meaningful if related to specific types of plants.

The authors acknowledge Mr. Beard's suggestion that a one-line diagram should be used in the survey of the electric utility supply. A new survey of the electric utility supply is being started, and Mr. Beard's suggestion will be included. This new survey should clear up the problem of the questionable accuracy mentioned by Mr. Beard. The authors acknowledge Mr. Beard's comment questioning the accuracy of the "time" factor in loads lost versus time of power outage in Table 30.

In answer to several questions raised by Mr. Krasnodebski, the authors make the following comments.

1) The failure rates are based upon a calendar year of 8760 h, not upon an operating time, which could be less and would thus result in a higher failure rate than reported in the survey.

2) The failures of circuit breakers are meant to include the auxiliaries.

3) The failure modes of circuit breakers are included in Table 41; this includes "fail to close," "fail to open," etc. However, data were not collected on the number of circuit breaker operations.

4) The Reliability Subcommittee does not consider that it would be appropriate for a technical society such as IEEE to collect and publish reliability data by name of manufacturer.

5) The authors agree that better record keeping of failures would improve survey results. It is expected that future surveys will cover only a few categories of electrical equipment that are considered trouble areas.

6) The authors acknowledge the logic in the very interesting comments made on synchronous motors and switchgear bus and generators.

7) The steam turbine generators in industrial plants probably have constant operation and thus could be expected to have a much lower failure rate than 60–89 MW units in utility applications where the operation was cyclical.

The authors wish to thank Mr. Kuehn for his suggestions in analyzing the data. These suggestions included 1) test hypothesis that part of data can be rejected, and 2) analysis of variance or multiple regression. Mr. Becker has raised a point where this approach for analyzing the data could possibly be tried. Mr. Becker feels that the survey results are too high on the downtime per failure of a single-circuit electric utility supply. This may be true for his system, but perhaps other utilities are not as good as his company's system.

Mr. Wong has raised a warning about drawing the conclusion that equipment reliability has improved since the previous survey conducted 11 to 12 years earlier. A separate paper has been prepared on this subject and will be published in the near future. This paper contains the conclusion that the failure rate of electrical equipment has shown a definite trend of improvement during the 12-year interval.

The authors wish to thank Mr. Wells for his discussion on preventive maintenance. A lot more data on preventive maintenance are being processed and will be included in Part 4. Mr. Wells' Table B shows more failures in the "12 months or more" category than for the "less than 12 months ago" category. The authors would like to point out that the electrical equipment has more unit-years of exposure in the "12 months or more" category and thus could be expected to have more failures. Thus it is not possible to conclude that more frequent preventive maintenance will reduce the failure rate. The Reliability Subcommittee is investigating this subject in further detail and will publish the results in Part 4.

Appendix B

Report on Reliability Survey of Industrial Plants

Part 4
Additional Detailed Tabulation of Some Data Previously Reported in the First Three Parts

Part 5
Plant Climate, Atmosphere, and Operating Schedule, the Average Age of Electrical Equipment, Percent Production Lost, and the Method of Restoring Electrical Service After a Failure

Part 6
Maintenance Quality of Electrical Equipment

By
Reliability Subcommittee
Industrial & Commercial Power Systems Committee
IEEE Industry Applications Society

A. D. Patton, ***Chairman***

C. E. Becker
W. H. Dickinson
P. E. Gannon
C. R. Heising
D. W. McWilliams
R. W. Parisian
S. Wells

Industrial and Commercial Power Systems Technical Conference
Institute of Electrical & Electronics Engineers, Inc
Denver, Colorado
June 3-6, 1974

Also Published
IEEE Transactions on Industry Applications
Jul/Aug and Sept/Oct 1974
Part 4 pp 456-462
Part 5 pp 463-466
Part 6 pp 467-468, 681, 469-476

Reprinted from pp 113-136
in 74CH0855-71A, the 1974
I&CPS Technical Conference

Report on Reliability Survey of Industrial Plants, Part IV: Additional Detailed Tabulation of Some Data Previously Reported in the First Three Parts

IEEE COMMITTEE REPORT

***Abstract*—An IEEE sponsored reliability survey of industrial plants was completed during 1972. This survey included 30 companies covering a total of 68 industrial plants in the United States and Canada. Additional detailed results are reported on some data that were previously reported in the first three parts. This includes failure modes of circuit breakers, cost of power outages, critical service loss duration time, loss of motor load versus time of power outage, and the effect of failure repair method and repair urgency on the average downtime per failure of electrical equipment. This information is useful in the design of industrial power distribution systems.**

INTRODUCTION AND RESULTS

During 1972 the Reliability Subcommittee of the Industrial and Commercial Power Systems Committee completed a reliability survey of industrial plants. This paper presents Part IV of the results from the survey. The first three parts [1]–[3] were published previously. Some of the data in the first three parts caused questions to be raised about the possibility of obtaining additional details. These additional details are being reported in this paper and include the following results.

Table 43 gives failure modes of circuit breakers, including

1) metalclad drawout
   a) 0–600 V
   b) 601–15 000 V
   c) all voltages
2) fixed type (includes molded case)
   a) 0–600 V
   b) all voltages.

Tables 44, 45 give cost of power outages, adding 25 and 75 percentile data to what was previously published.

Table 46 gives critical service loss duration time (maximum length of power failure that will not stop plant production), adding 10, 25, 75, and 90 percentile data to what was previously published.

Paper TOD-74-33, approved by the Industrial and Commercial Power Systems Committee of the IEEE Industry Applications Society for presentation at the 1974 Industrial and Commercial Power Systems Technical Conference, Denver, Colo., June 2–6. Manuscript released for publication April 15, 1974.

Members of the Reliability Subcommittee of the IEEE Industrial and Commercial Power Systems Committee are A. D. Patton, *Chairman*, C. E. Becker, W. H. Dickinson, P. E. Gannon, C. R. Heising, D. W. McWilliams, R. W. Parisian, and S. Wells.

Table 47 lists loss of motor load versus time of power outage, adding the following length of power outage categories:

1) 10 to 15 cycles
2) 15+ to 30 cycles
3) 0.5+ to 2.0 s
4) 2+ to 4.0 s
5) >4.0 s.

Tables 48 through 56 report the effect of failure repair method and failure repair urgency on the average downtime per failure for the following equipment categories:

1) transformers—liquid filled
   a) 601–15 000 V
   b) above 15 000 V
2) circuit breakers—metalclad drawout
   a) 0–600 V
   b) above 600 V
3) motors
   a) induction, 0–600 V
   b) induction, 601–15 000 V
   c) synchronous, 601–15 000 V
4) cable
   a) above ground and aerial, 601–15 000 V
   b) below ground and direct burial, 601–15 000 V

In each of the Tables 43 through 56 reference is made to the tables in Parts I, II, and III where previous results had been reported.

DISCUSSION—FAILURE MODES OF CIRCUIT BREAKERS

The data on failure modes of circuit breakers given in Table 43 show some very interesting results.

Circuit Breakers, 0–600 V

71 percent of the failures of metalclad drawout circuit breakers were "opened when it shouldn't" versus 5 percent of the failures for fixed-type circuit breakers (includes molded case). 77 percent of the failures of fixed-type circuit breakers (includes molded case) were "failed while operating (not while opening or closing)," and only 10 percent of the metalclad drawout failures included this failure mode.

None of the failures reported for either type of circuit breaker were "failed while opening." Only 9 percent and

TABLE 43 - FAILURE MODES OF CIRCUIT BREAKERS - Percent of Total Failures in Each Failure Mode
(Data Previously Reported in Tables 5 and 41)

All Circuit Breakers	Metalclad Drawout-All	Metalclad Drawout-601-15,000 Volts	Metalclad Drawout-0-600 Volts All Sizes	*Fixed Type 0-600 Volts All Sizes	*Fixed Type-All	Card-Type 3, Col. 46
%	%	%	%	%	%	FAILURE CHARACTERISTIC
5	5	2	7	8	6	Failed to close when it should
9	12	21	0	0	2	Failed while opening
42	58	49	71	5	4	Opened when it shouldn't
7	6	4	9	5	4	Damaged while successfully opening
2	1	0	0	0	4	Damaged while closing
32	16	24	10	77	73	Failed while operating (not while opening or closing)
1	0	0	0	0	2	Failed during testing or maintenance
1	2	0	3	0	0	Damage discovered during testing or maintenance
1	0	0	0	5	5	Other
100%	100%	100%	100%	100%	100%	Total Percent
-	117	53	59	39	48	Number of Failures in Total Percent
-	7	0	7	1	1	Number Not Reported in Col. 46, Card-Type 3
-	124	-	66	40	49	Total Failures in Table 5

*Includes molded case

5 percent, respectively, of the failures were "damaged while successfully opening." Only 7 to 8 percent of the failures were "failed to close when it should."

It appears that the dominate failure mode for metalclad drawout circuit breakers, 0–600 V, is "opened when it shouldn't." It is possible that some of these failures were external to the breaker and of unknown cause and were blamed on the breaker. Some of these may have been due to improper setting of the trip current.

The dominate failure mode for fixed-type circuit breakers (includes molded case), 0–600 V, is "failed while operating (not while opening or closing)."

Metalclad Drawout Circuit Breakers, 601–15 000 V

Metal drawout circuit breakers, 601–15 000 V, had 21 percent of the failures classified as "failed while opening" and 4 percent classified as "damaged while successfully opening." Another 24 percent of the failures were classified as "failed while operating (not while opening or closing)." 49 percent of the failures were classified as "opened when it shouldn't;" it is suspected that some of these may have been due to improper setting of the trip current.

It appears that metalclad drawout circuit breakers, 601–15 000 V, have about half of their failures as "opened when it shouldn't" and the other half as "failed while operating or while opening."

DISCUSSION—LOSS OF MOTOR LOAD VERSUS TIME OF POWER OUTAGE

The data on loss of motor load shown in Table 47 indicate that for power outages greater than 10 cycles duration most of the plants lose the motor load. However, for power outages between 1 to 10 cycles duration, only about half as many lose the motor load. Thus, power outages of less than 10 cycles duration may often not result in losing the motor load.

There were many power outages of more than 4.0 s duration, and 35 percent did not lose motor load. It is suspected that many of these did not have a motor load. Some may have had a duplicate feed and thus did not lose the motor load.

DISCUSSION—EFFECT OF FAILURE REPAIR METHOD AND FAILURE REPAIR URGENCY ON AVERAGE HOURS DOWNTIME PER FAILURE

Data were given in Part I on the average hours downtime per failure for 74 categories of electrical equipment. It is known that the downtime after a failure can be affected to a large extent by the failure repair method and the failure repair urgency. The failure repair method includes either repair of the failed component or else replacement with a spare. Some data were given in Tables 33 and 34 of Part III on the failure repair method and the failure repair urgency for whole classes of electrical equipment.

A more detailed study is reported in Tables 48–56 of this paper on the effect of the failure repair method and the failure repair urgency on the average hours downtime per failure. This is only reported for 9 electrical equipment categories, rather than the 74 categories given in Part I. These 9 electrical equipment categories were selected because an adequate sample size existed of the number of failures and because the average downtime per failure was effected significantly by the failure repair method and/or the failure repair urgency.

TABLE 44 - PLANT OUTAGE COST PER FAILURE PER kW OF MAXIMUM DEMAND ($ per kW)
All Industry - USA & Canada
(Data Previously Reported in Tables 22, 24 and 26)

Plant Size	Number of Plants Reporting	Minimum	25% Percentile	Median	75% Percentile	Maximum	Average
All Plants	42	.002	.17	.69	2.55	10.00	1.89
Plants > 1000 kW Max. Demand	32	.002	.09	.32	1.31	7.50	1.05
Plants < 1000 kW Max. Demand	10	.50	1.71	3.68	8.27	10.00	4.59

TABLE 45 - PLANT OUTAGE COST PER HR. DOWNTIME PER kW OF MAXIMUM DEMAND ($ per kWh)
All Industry - USA & Canada
(Data Previously Reported in Tables 23, 25 and 27)

Plant Size	Number of Plants Reporting	Minimum	25% Percentile	Median	75% Percentile	Maximum	Average
All Plants	41	.0009	.18	.83	2.71	27.00	2.68
Plants > 1000 kW Max. Demand	31	.0009	.12	.36	1.20	5.77	.94
Plants < 1000 kW Max. Demand	10	.86	1.83	4.42	12.50	27.00	8.11

TABLE 46 - CRITICAL SERVICE LOSS DURATION (Maximum Length of Power Failure that Will Not Stop Plant Production)
(Data Previously Reported in Table 29)

Industry	Number of Plants Reporting	Average	10% Percentile	25% Percentile	Median	75% Percentile	90% Percentile
All Industry - USA & Canada	55	12.6 min.	5.0 cycles	10.0 cycles	10.0 sec.	15.0 min.	60.0 min.
Chemical	20	4.56 min.	3.2 cycles	8.5 cycles	1.25 sec.	5.0 min.	28.5 min.

TABLE 47 - LOSS OF MOTOR LOAD VERSUS TIME OF POWER OUTAGE
Tabulation of the Percentage of Equipment Failures for Which the Motor Load was Lost
(Data Previously Reported in Table 30)

Length of Equipment Failure	Number of Failures Reported	TYPE OF LOAD LOST		
		Motor		
		Yes	No	Not Known
1 cycle or less	0	0%	0%	0%
1+ to 10- cycles	-	33%	67%	0%
10 to 15 cycles	7	86%	14%	0%
15+ to 30 cycles	28	96%	4%	0%
0.5+ to 2.0 sec.	30	77%	13%	10%
2.0+ to 4.0 sec.	10	100%	0%	0%
>4.0 second	998	64%	35%	0%

TABLE 48 TRANSFORMERS-LIQUID FILLED, 601-15,000 VOLTS
EFFECT OF FAILURE REPAIR METHOD AND FAILURE REPAIR URGENCY
ON THE AVERAGE HOURS DOWNTIME PER FAILURE
(Previous Data Given in Tables 4, 33 and 34)

FAILURE REPAIR METHOD			FAILURE REPAIR METHOD		
Repair	Replace with Spare	Total	Repair	Replace with Spare	
Number of Failures			Average Hours Downtime per Failure		FAILURE REPAIR URGENCY
4	22	26	*	130	1. Requiring round-the-clock all out efforts
10	3	13	342	*	2. Requiring repair work only during regular workday, perhaps with some overtime
0	0	0	-	-	3. Requiring repair work on a non-priority basis
14	25	39	Average 174. Hours		Total

*Small Sample Size

TABLE 49 - TRANSFORMERS-LIQUID FILLED, ABOVE 15,000 VOLTS
EFFECT OF FAILURE REPAIR METHOD AND FAILURE REPAIR URGENCY
ON THE AVERAGE HOURS DOWNTIME PER FAILURE
(Previous Data Given in Tables 4, 33 and 34)

FAILURE REPAIR METHOD			FAILURE REPAIR METHOD		
Repair	Replace with Spare	Total	Repair	Replace with Spare	
Number of Failures			Average Hours Downtime per Failure		FAILURE REPAIR URGENCY
2	5	7	*	*	1. Requiring round-the-clock all out efforts
12	4	16	1842	*	2. Requiring repair work only during regular workday, perhaps with some overtime
0	1	1	-	*	3. Requiring repair work on a non-priority basis
14	10	24	Average 1076. Hours		Total

*Small Sample Size

TABLE 50 - CIRCUIT BREAKERS - METALCLAD DRAWOUT, 0-600 VOLTS
EFFECT OF FAILURE REPAIR METHOD AND FAILURE REPAIR URGENCY
ON THE AVERAGE HOURS DOWNTIME PER FAILURE
(Previous Data Given in Tables 5, 33 and 34)

FAILURE REPAIR METHOD			FAILURE REPAIR METHOD		
Repair	Replace with Spare	Total	Repair	Replace with Spare	
Number of Failures			Average Hours Downtime per Failure		FAILURE REPAIR URGENCY
31	19	50	3.3	3.8	1. Requiring round-the-clock all out efforts
6	1	7	*	*	2. Requiring repair work only during regular workday, perhaps with some overtime
8	1	9	*	*	3. Requiring repair work on a non-priority basis
45	21	66	Average 147. Hours		Total

*Small Sample Size

TABLE 51 - CIRCUIT BREAKERS - METALCLAD DRAWOUT, ABOVE 600 VOLTS
EFFECT OF FAILURE REPAIR METHOD AND FAILURE REPAIR URGENCY
ON THE AVERAGE HOURS DOWNTIME PER FAILURE
(Previous Data Given in Tables 5, 33 and 34)

FAILURE REPAIR METHOD			FAILURE REPAIR METHOD		
Repair	Replace with Spare	Total	Repair	Replace with Spare	
Number of Failures			Average Hours Downtime per Failure		FAILURE REPAIR URGENCY
34	12	46	83.1	2.1	1. Requiring round-the-clock all out efforts
3	9	12	*	*	2. Requiring repair work only during regular workday, perhaps with some overtime
0	0	0	-	-	3. Requiring repair work on a non-priority basis
37	21	58	Average 109. Hours		Total

*Small Sample Size

TABLE 52 - MOTORS - INDUCTION, 0-600 VOLTS
EFFECT OF FAILURE REPAIR METHOD AND FAILURE REPAIR URGENCY
ON THE AVERAGE HOURS DOWNTIME PER FAILURE
(Previous Data Given in Tables 7, 33 and 34)

FAILURE REPAIR METHOD			FAILURE REPAIR METHOD		
Repair	Replace with Spare	Total	Repair	Replace with Spare	
Number of Failures			Average Hours Downtime per Failure		FAILURE REPAIR URGENCY
12	19	31	44.7	6.6	1. Requiring round-the-clock all out efforts
175	2	177	123	*	2. Requiring repair work only during regular workday, perhaps with some overtime
0	5	5	-	*	3. Requiring repair work on a non-priority basis
187	26	213	Average 114. Hours		Total

*Small Sample Size

TABLE 53 - MOTORS - INDUCTION, 601-15,000 VOLTS
EFFECT OF FAILURE REPAIR METHOD AND FAILURE REPAIR URGENCY
ON THE AVERAGE HOURS DOWNTIME PER FAILURE
(Previous Data Given in Tables 7, 33 and 34)

FAILURE REPAIR METHOD			FAILURE REPAIR METHOD		
Repair	Replace with Spare	Total	Repair	Replace with Spare	
Number of Failures			Average Hours Downtime per Failure		FAILURE REPAIR URGENCY
14	10	24	88.1	*	1. Requiring round-the-clock all out efforts
93	48	141	83.6	34.7	2. Requiring repair work only during regular workday, perhaps with some overtime
6	0	6	*	-	3. Requiring repair work on a non-priority basis
113	58	171	Average 76. Hours		Total

*Small Sample Size

TABLE 54 - MOTORS - SYNCHRONOUS, 601 - 15,000 VOLTS
EFFECT OF FAILURE REPAIR METHOD AND FAILURE REPAIR URGENCY
ON THE AVERAGE HOURS DOWNTIME PER FAILURE
(Previous Data Given in Tables 7, 33 and 34)

FAILURE REPAIR METHOD			FAILURE REPAIR METHOD		
Repair	Replace with Spare	Total	Repair	Replace with Spare	
Number of Failures			Average Hours Downtime per Failure		FAILURE REPAIR URGENCY
28	2	30	198	*	1. Requiring round-the-clock all out efforts
55	8	63	201	*	2. Requiring repair work only during regular workday, perhaps with some overtime
1	0	1	*	-	3. Requiring repair work on a non-priority basis
84	10	94	Average 175. Hours		Total

*Small Sample Size

TABLE 55 - CABLE - ABOVE GROUND & AERIAL, 601-15,000 VOLTS
EFFECT OF FAILURE REPAIR METHOD AND FAILURE REPAIR URGENCY
ON THE AVERAGE HOURS DOWNTIME PER FAILURE
(Previous Data Given in Tables 13, 33 and 34)

FAILURE REPAIR METHOD			FAILURE REPAIR METHOD		
Repair	Replace with Spare	Total	Repair	Replace with Spare	
Number of Failures			Average Hours Downtime per Failure		FAILURE REPAIR URGENCY
46	4	50	9.0	*	1. Requiring round-the-clock all out efforts
11	8	19	*	*	2. Requiring repair work only during regular workday, perhaps with some overtime
2	2	4	*	*	3. Requiring repair work on a non-priority basis
59	14	73	Average 40.4 Hours		Total

*Small Sample Size

TABLE 56 - CABLE - BELOW GROUND & DIRECT BURIAL, 601-15,000 VOLTS
EFFECT OF FAILURE REPAIR METHOD AND FAILURE REPAIR URGENCY
ON THE AVERAGE HOURS DOWNTIME PER FAILURE
(Previous Data Given in Tables 13, 33 and 34)

FAILURE REPAIR METHOD			FAILURE REPAIR METHOD		
Repair	Replace with Spare	Total	Repair	Replace with Spare	
Number of Failures			Average Hours Downtime per Failure		FAILURE REPAIR URGENCY
17	57	74	26.5	19.0	1. Requiring round-the-clock all out efforts
2	33	35	*	77.8	2. Requiring repair work only during regular workday, perhaps with some overtime
3	3	6	*	*	3. Requiring repair work on a non-priority basis
22	93	115	Average 95.5 Hours		Total

*Small Sample Size

In several cases there is a disparity in the downtime between the "average" and the cases where work is done "round the clock." When making availability calculations, this should be considered when deciding what value to use for the downtime after a failure.

Transformers, Liquid Filled

Transformers, above 15 000 V, had an average downtime per failure of 1842 h when sent out for repair without round-the-clock urgency. This compares with an overall average of 1076 h for all outage times, which included several cases of replacement with a spare. Thus it can be concluded that repair gives a much longer outage time than replacement with a spare for transformers, above 15 000 V.

Transformers, 601–15 000 V, had an average downtime per failure of 342 h when sent out for repair without round-the-clock urgency. This compares with 130 h for replacement with a spare while working round the clock. Thus it can be concluded that repair gives a much longer outage time for transformers, 601–15 000 V, than replacement with a spare while working round the clock.

Circuit Breakers, Metalclad Drawout

Metalclad drawout circuit breakers, 0–600 V, had an average downtime per failure of 3.3 h to 3.8 h when fixing the failure with round-the-clock efforts. This compares with an overall average of 147 h for all outage times. Thus it can be concluded that 24 percent of the outages of metalclad drawout circuit breakers, 0–600 V, had low urgency for fixing the failure, and that these 24 percent of the failures resulted in increasing the average downtime per failure from 3.8 h to 147 h.

Metalclad drawout circuit breakers above 600 V, had an average downtime per failure of 109 h for all outages. However, when round-the-clock effort was applied it only took 83 h for repair and only took 2.1 h for replacement with a spare. This shows that it is possible to reduce the downtime by having a spare and working round the clock when fixing metalclad drawout circuit breakers, above 600 V.

Motors

Most users of synchronous motors, 601–15 000 V, did not have a spare. Thus the average downtime per failure was 175 h for all failures.

Induction motors, 601–15 000 V, had an average downtime per failure of 35 h for replacement with a spare, compared to 84 to 88 h for repair. Induction motors, 0–600 V, had an average downtime per failure of 6.6 h for replacement with a spare while working round the clock. This compares with 123 h for repair and not working round the clock.

Cables

Cables, above ground and aerial, 601–15 000 V, had an average downtime per failure of 9 h for repair when working round the clock. This compares with 40 h for all failures. This shows that it is possible to reduce the downtime by working round the clock when fixing cables, above ground and aerial, 601–15 000 V.

Cables, below ground and direct burial, 601–15 000 V, had an average downtime per failure of 96 h for all failures. However, this was only 19 to 27 h when working round the clock. This shows that it is possible to reduce the downtime by working round the clock when fixing cables, below ground and direct burial, 601–15 000 V.

DISCUSSION—COST OF POWER OUTAGES

Data are given in Tables 44 and 45 on the cost of power outages to industrial plants. This has added 25th and 75th percentile data to what had previously been reported in Part II. These were added because of the wide spread in the cost of power outages to industrial plants.

REFERENCES

[1] W. H. Dickinson *et al.*, "Report on reliability survey of industrial plants, part I: Reliability of electrical equipment," *IEEE Trans. Ind. Appl.*, vol. IA-10, pp. 213–235, Mar./Apr. 1974.
[2] W. H. Dickinson *et al.*, "Report on reliability survey of industrial plants, part II: Cost of power outages, plant restart time, critical service loss duration time, and type of loads lost versus time of power outages," *IEEE Trans. Ind. Appl.*, vol. IA-10, pp. 236–241, Mar./Apr. 1974.
[3] W. H. Dickinson *et al.*, "Report on reliability survey of industrial plants, part III: Causes and types of failures of electrical equipment, the methods of repair, and the urgency of repair," *IEEE Trans. Ind. Appl.*, vol. 14-10, pp. 242–252, Mar./Apr. 1974.

Report on Reliability Survey of Industrial Plants, Part V: Plant Climate, Atmosphere, and Operating Schedule, the Average Age of Electrical Equipment, Percent Production Lost, and the Method of Restoring Electrical Service after a Failure

IEEE COMMITTEE REPORT

Abstract—**An IEEE sponsored reliability survey of industrial plants was completed during 1972. This survey included the plant climate, atmosphere, and operating schedule, the average age of electrical equipment, percent production lost, and the method of restoring electrical service after a failure. The results are reported from the survey of 30 companies covering 68 plants in nine industries in the United States and Canada. This information is useful in the design of industrial power distribution systems.**

INTRODUCTION AND RESULTS

DURING 1972 the Reliability Subcommittee of the Industrial and Commercial Power Systems Committee completed a reliability survey of industrial plants. This paper presents Part V of the results from the survey. The first three parts [1]–[3] were published previously; some of the data of lesser importance were not published at that time but are presented in this paper. Included in Part V are

- Table 57—Failure Forewarning for Public Utility Power Interruption Only,
- Table 58—Percent Production Lost,
- Table 59—Method of Service Restoration,
- Table 60—Average Age of Electrical Equipment,
- Table 61—Plant Climate,
- Table 62—Plant Atmosphere,
- Table 63—Plant Operating Schedule.

These data are useful when using the results published in Parts I, II, III, IV [4], and VI [5]. This information is also useful in the design of industrial power distribution systems. The data on average age of electrical equipment and plant operating schedule provide answers to some points raised in the written discussion to Part I.

Paper TOD-74-33, approved by the Industrial and Commercial Power Systems Committee of the IEEE Industry Applications Society for presentation at the 1974 Industrial and Commercial Power Systems Technical Conference, Denver, Colo., June 2–6. Manuscript released for publication April 15, 1974.

Members of the Reliability Subcommittee of the IEEE Industrial and Commercial Power Systems Committee are A. D. Patton, *Chairman*, C. E. Becker, W. H. Dickinson, P. E. Gannon, C. R. Heising, D. W. McWilliams, R. W. Parisian, and S. Wells.

TABLE 57 - FAILURE FOREWARNING for PUBLIC UTILITY POWER INTERRUPTION ONLY

Percent	Col. 25 Card-Type 3
97%	1. If no forewarning was given
3%	2. If forewarning was given
—	For other types of failure, leave blank
100%	Total Percent
172	Total Interruptions Reported

SURVEY FORM

The survey form is shown in Appendix A of Part I [1]. The information reported in this paper came from 1) card type 3, columns 25, 53, and 58; 2) card type 2, column 33; and 3) card type 1, columns 9–11 and 13. The definition of *failure* is given in Part I.

RESPONSE TO SURVEY

A total of 30 companies responded to the survey questionnaire, reporting data covering 68 plants in nine industries in the United States and Canada. For the purpose of reporting results in this paper, Part V, the number of industries were reduced from nine down to five plus an "all other" category. The five industries selected were the ones for which equipment failure rate data were reported in Tables 3 through 19, Part I. All of the remaining industries were combined into an "all other" category in Tables 61–63 on plant climate, plant atmosphere, and plant operating schedule.

DISCUSSION—FOREWARNING FOR PUBLIC UTILITY POWER INTERRUPTION

Only 3 percent of the time was a failure forewarning given for a public utility power interruption to the industrial plant. Data from Table 3, Part I, and Table 57, Part V, indicate that a large percentage of these interruptions were on double- or triple-circuit supplies. Forewarning can be important to plants with a single circuit. It can also be important to plants containing a double circuit with manual switchover.

TABLE 58 - PERCENT PRODUCTION LOST

Col. 53 Card Type 3 Percent Production Lost	ELECTRIC UTILITY POWER SUPPLIES	TRANSFORMERS	CIRCUIT BREAKERS	MOTOR STARTERS	MOTORS	GENERATORS	DISCONNECT SWITCHES	SWITCHGEAR BUS-INSULATED	SWITCHGEAR BUS-BARE	BUS DUCT	OPEN WIRE	CABLE	CABLE JOINTS	CABLE TERMINATIONS
	%	%	%	%	%	%	%	%	%	%	%	%	%	%
0 None	41	22	19	85	24	80	20	20	17	30	62	28	33	47
1 0-30 Percent	32	63	73	13	73	5	75	60	33	55	25	60	58	35
2 Above 30 Percent	27	15	8	2	3	15	5	20	50	15	13	13	9	18
Total Percent	100	100	100	100	100	100	100	100	100	100	100	101	100	100
Total Failures Reported	202	101	177	168	561	85	101	20	24	20	108	223	45	51

TABLE 59 - METHOD OF SERVICE RESTORATION

TOTAL	ELECTRIC UTILITY POWER SUPPLIES	TRANSFORMERS	CIRCUIT BREAKERS	MOTOR STARTERS	MOTORS	GENERATORS	DISCONNECT SWITCHES	SWITCHGEAR BUS - INSULATED	SWITCHGEAR BUS - BARE	BUS DUCT	OPEN WIRE	CABLE	CABLE JOINTS	CABLE TERMINATIONS	Col. 58 Card Type 3
%	%	%	%	%	%	%	%	%	%	%	%	%	%	%	Give method of restoring service to plant
7	1	3	6	0	5	20	0	58	25	20	13	14	28	19	1 Primary selection - manual
2	8	0	1	0	0	0	0	0	5	0	4	5	8	0	2 Primary selection - automatic
11	1	25	6	0	14	33	0	17	10	10	2	20	32	23	3 Secondary selection - manual
2	1	3	8	0	0	0	0	0	0	0	1	0	8	4	4 Secondary selection - automatic
0+	0	0	0	0	0	0	0	0	5	0	0	0	0	0	5 Network protector operation - automatic
22	5	25	11	12	30	20	3	17	20	35	31	42	24	27	6 Repair of failed component
22	2	39	38	10	29	14	77	0	10	35	6	2	0	12	7 Replacement of failed component with spare
12	81	0	1	0	0	13	0	0	0	0	1	1	0	0	8 Utility restored service
22	1	5	29	78	22	0	20	8	25	0	42	16	0	15	9 Other - explain in remarks
100	100	100	100	100	100	100	100	100	100	100	100	100	100	100	Total Percent
1204	171	75	160	68	318	15	69	12	20	20	103	122	25	26	TOTAL NUMBER REPORTED

TABLE 60 - AVERAGE AGE OF ELECTRICAL EQUIPMENT

TRANSFORMERS	CIRCUIT BREAKERS	MOTOR STARTERS	MOTORS	GENERATORS	DISCONNECT SWITCHES	SWITCHGEAR BUS-INSULATED	SWITCHGEAR BUS-BARE	BUS DUCT	OPEN WIRE	CABLE	CABLE JOINTS	CABLE TERMINATIONS	
NUMBER OF INSTALLED UNITS													Age
6	989	101	104	0	0	0	0	0	30	15	0	12	1 Less than 1 year old
694	3691	3162	1884	9	909	646	1998	1206	12	1019	1385	3314	2 1-10 years od
835	1944	608	3643	77	552	691	555	13640	472	1831	2338	5712	3 More than 10 years old

TABLE 61 - PLANT CLIMATE (for entire plant site)
TABLE 62 - PLANT ATMOSPHERE (for entire plant site)

ALL INDUSTRY	CHEMICAL	METAL	PETROLEUM	RUBBER AND PLASTICS	TEXTILE	ALL OTHER	Table, Title, Card-Type 1 Column No.	
NUMBER OF PLANTS							TABLE 61 - PLANT CLIMATE (Col. 9)	
							Average of Daily Maximums for Hottest Month	
							Temperature	Relative Humidity (RH) (measured at noon to 2 PM ST)
14	8	1	3	0	1	1	1 Hot (> 90F)	High (> 55 RH)
3	3	0	0	0	0	0	2 Hot (> 90F)	Moderate (50-55 RH)
12	0	0	0	0	0	12	3 Hot (> 90F)	Low (< 50 RH)
14	4	1	2	0	0	7	4 Moderate (80-90F)	High (> 55 RH)
16	5	1	0	1	1	8	5 Moderate (80-90F)	Moderate (50-55RH)
6	1	0	1	2	1	1	6 Moderate (80-90F)	Low (< 50 RH)
1	0	0	0	0	0	1	7 Low (<80F)	High (> 55 RH)
2	0	0	2	0	0	0	8 Low (< 80F)	Moderate (50-55 RH)
0	0	0	0	0	0	0	9 Low (< 80F)	Low (< 50 RH)
							TABLE 62 - PLANT ATMOSPHERE (Col. 10)	
34	2	1	7	0	2	22	1 Clean to slightly polluted air	
5	4	0	1	0	0	0	2 With salt spray and corrosive chemicals	
0	0	0	0	0	0	0	3 With salt spray and dust or sand	
0	0	0	0	0	0	0	4 With salt spray only	
13	8	0	0	1	1	3	5 With corrosive chemicals and dust or sand	
4	4	0	0	0	0	0	6 With corrosive chemicals only	
2	0	0	0	0	0	2	7 With dust or sand only	
5	0	2	0	2	0	1	8 With conductive dust	
1	0	0	0	0	0	1	9 Other	

TABLE 63 - PLANT OPERATING SCHEDULE

ALL INDUSTRY	CHEMICAL	METAL	PETROLEUM	RUBBER AND PLASTICS	TEXTILE	ALL OTHER	Title, Card-Type 1 Column No.
NUMBER OF PLANTS							HOURS PER DAY (Col. 11)
0	0	0	0	0	0	0	Less than 8
9	2	0	1	0	0	6	8
0	0	0	0	0	0	0	9 to 15
19	0	2	0	0	0	17	16
0	0	0	0	0	0	0	17 to 23
40	19	1	7	3	3	7	24
							DAYS PER WEEK (Col. 13)
0	0	0	0	0	0	0	Less than 5
30	1	2	1	2	0	24	5
3	1	0	0	0	0	2	6
35	19	1	7	1	3	4	7

DISCUSSION—PERCENT PRODUCTION LOST

The most severe category of failure in an industrial plant is where above 30 percent of the production is lost. Data from Table 58 show that the following percent of equipment class failures resulted in losing above 30 percent of the production.

Switchgear bus—bare	50 percent
Electric utility power supplies	27 percent
Switchgear bus—insulated	20 percent
Cable terminations	18 percent
Bus duct	15 percent
Transformers	15 percent
Generators	15 percent
Open wire	13 percent
Cable	13 percent
Cable joints	9 percent
Circuit breakers	8 percent
Motors	3 percent
Motor starters	2 percent

It can be seen that failures of switchgear bus and electric utility power supplies often result in losing above 30 percent of the production.

DISCUSSION—METHOD OF SERVICE RESTORATION

The data on method of electrical service restoration to plant is shown in Table 59. A percentage breakdown of the total shows the following results.

Replacement of failed component with spare	22 percent
Repair of failed components	22 percent
Other	22 percent
Utility service restored	12 percent
Secondary selection—manual	11 percent
Primary selection—manual	7 percent
Primary selection—automatic	2 percent
Secondary selection—automatic	2 percent
Network protector operation—automatic	0+ percent

The most common methods of service restoration are replacement of failed component with a spare or repair of failed component. Only 22 percent of the time is primary selection or secondary selection used; this would indicate that most power distribution systems are radial.

DISCUSSION—AVERAGE AGE OF ELECTRICAL EQUIPMENT

Many respondents to the reliability survey of industrial plants submitted data covering a ten-year period. Thus it is not surprising to see that Table 60 shows a large population that is more than ten years old. The following percent of installed units are classified as more than ten years old.

Bus duct	92 percent
Open wire	92 percent
Generators	90 percent
Motors	65 percent
Cable	64 percent
Cable joints	63 percent
Cable terminations	63 percent
Transformers	54 percent
Switchgear bus—insulated	52 percent

Motor starters, disconnect switches, switchgear bus—bare, and circuit breakers had over 50 percent of the installed units one to ten years old.

15 percent of the circuit breakers were less than one year old. All other equipment classes had less than 6 percent of the installed units less than a year old.

DISCUSSION—PLANT CLIMATE AND ATMOSPHERE

Data on plant climate and plant atmosphere are given in Tables 61 and 62. 43 percent of the plants were in a hot climate, 53 percent in a moderate climate, and only 4 percent in a low climate (cold climate). 43 percent of the plants had high relative humidity, 31 percent had moderate relative humidity, and 26 percent had low relative humidity. 53 percent of the plants had a plant atmosphere classified as "clean to slightly polluted air." The other 47 percent had an atmosphere with some contamination.

DISCUSSION—PLANT OPERATING SCHEDULE

The data on plant operating schedule are given in Table 63. 52 percent of the plants operated 7 days per week, 4 percent for 6 days, and 44 percent for 5 days. 59 percent of the plants operated 24 h per day, 28 percent for 16 h, and 13 percent for 8 h.

REFERENCES

[1] W. H. Dickinson *et al.*, "Report on reliability survey of industrial plants, part I: Reliability of electrical equipment," *IEEE Trans. Ind. Appl.*, vol. IA-10, pp. 213–235, Mar./Apr. 1974.

[2] W. H. Dickinson *et al.*, "Report on reliability survey of industrial plants, part II: Cost of power outages, plant restart time, critical service loss duration time, and type of loads lost versus time of power outages," *IEEE Trans. Ind. Appl.*, vol. IA-10, pp. 236–241, Mar./Apr. 1974.

[3] W. H. Dickinson *et al.*, "Report on reliability survey of industrial plants, part III: Causes and types of failures of electrical equipment, the methods of repair, and the urgency of repair," *IEEE Trans. Ind. Appl.*, vol. IA-10, pp. 242–249, Mar./Apr. 1974.

[4] A. D. Patton *et al.*, "Report on reliability survey of industrial plants, part IV: Additional detailed tabulation of some data previously reported in the first three parts," this issue, pp. 456–462.

[5] A. D. Patton *et al.*, "Report on reliability survey of industrial plants, part VI: Maintenance quality of electrical equipment," this issue, pp. 467–476.

Report on Reliability Survey of Industrial Plants, Part VI: Maintenance Quality of Electrical Equipment

IEEE COMMITTEE REPORT

***Abstract*—An IEEE sponsored reliability survey of industrial plants was completed during 1972. This included maintenance quality, the frequency of schedule maintenance, and the failures caused by inadequate maintenance. The results are reported from the survey of 30 companies covering 68 plants in nine industries in the United States and Canada. This information is useful in the design of industrial power distribution systems.**

INTRODUCTION

A KNOWLEDGE of maintenance quality of electrical equipment in industrial plants is useful information when planning the maintenance program of industrial power distribution systems. In addition it is useful to know how this correlates with the normal maintenance cycle and the failures blamed on inadequate maintenance. During 1972 the Reliability Subcommittee of the Industrial and Commercial Power Systems Committee completed a reliability survey of industrial plants. This paper presents Part VI of the results from the survey. The first three parts [1]–[3] were published previously. Table 38 from Part III reported that inadequate maintenance was blamed for between 8 to 30 percent of the failures of electrical equipment. This information has caused the Reliability Subcommittee to make a further study of the failure data; the results from this study are being reported in this paper. Included in Part VI are the results for 12 main classes of electrical equipment on

1) equipment population versus a) maintenance quality and b) normal maintenance cycle;
2) failures due to all causes versus a) failure, months since maintained, and b) maintenance quality;
3) failures due to inadequate maintenance versus a) failure, months since maintained, and b) maintenance quality.

The "maintenance quality" is an opinion that was reported by each participant in the survey. The four classifications used were "excellent," "fair," "poor," and "none." The "normal maintenance" cycle is the frequency of performing preventive maintenance.

Paper TOD-74-33, approved by the Industrial and Commercial Power Systems Committee of the IEEE Industry Applications Society for presentation at the 1974 Industrial and Commercial Power Systems Technical Conference, Denver, Colo., June 2–6. Manuscript released for publication April 15, 1974.

Members of the Reliability Subcommittee of the IEEE Industrial and Commercial Power Systems Committee are A. D. Patton, *Chairman*, C. E. Becker, W. H. Dickinson, P. E. Gannon, C. R. Heising, D. W. McWilliams, R. W. Parisian, and S. Wells.

SURVEY FORM

The survey form is shown in Appendix A of Part I [1]. The information reported in this paper came from 1) card type 2, col. 34 (maintenance, normal cycle); 2) card type 2, col. 36 (maintenance quality); 3) card type 3, col. 34 (failure, months since maintained); 4) card type 3, col. 40 (suspected failure responsibility). The definition of *failure* is given in Part I.

RESPONSE TO SURVEY

A total of 30 companies responded to the survey questionnaire, reporting data from nine industries in the United States and Canada. Every plant did not report all the information called for in card type 2, columns 34 and 36. Every failure report did not have filled out all of the information called for in card type 3, columns 34 and 40; a total of 1469 failures had this information filled in and are reported here in Part VI, and 240 of these failures were blamed on inadequate maintenance. Differences in the number of failures and unit-years reported here in Part VI and those previously reported in Part I and Part III can be explained from the preceding.

STATISTICAL ANALYSIS

The subject of statistical analysis of equipment failures is discussed in Part I [1]. Confidence limits for the failure rate are shown in Fig. 1 of Part I. The Reliability Subcommittee concluded that eight failures is an adequate sample size for reporting failure rates in the summary in Table 2, Part I. In a few cases, failure rate data were reported in Tables 3 through 19, Part I, where there were less than eight failures.

In this paper several cases are reported in Tables 67 through 78, where the failure rate contains less than eight failures; these cases have been marked "small sample size."

SURVEY RESULTS

Results are tabulated for 12 main equipment classes in Table 64 where the equipment population is given versus 1) maintenance quality and 2) normal maintenance cycle.

Table 65 summarizes the percent of each electrical equipment class population versus the maintenance quality. Table 66 summarizes the percent of each electrical equipment class population versus the normal maintenance cycle.

Results are tabulated for each of the 12 main equipment classes in Tables 67 through 78, where the number of failures is given for 1) failures due to all causes and 2)

Correction to "Report on Reliability Survey of Industrial Plants, Part VI: Maintenance Quality of Electrical Equipment"

IEEE COMMITTEE REPORT

TABLE 64 - POPULATION OF ELECTRICAL EQUIPMENT VERSUS MAINTENANCE QUALITY & NORMAL MAINTENANCE CYCLE

MAINTENANCE QUALITY Card-Type 2 Col. 36	MAINTENANCE, NORMAL CYCLE Card-Type 2 Col. 34 — Less Than 12 Months	12 - 24 Months	More Than 24 Months	No Preventive Maintenance	Total
	Population: Unit-Years				
TRANSFORMERS					
Excellent	19	8,904	2,314	0	11,237
Fair	292	3,081	5,961	0	9,334
Poor	0	130	210	0	340
None	0	0	0	39	39
Total	311	12,115	8,485	39	20,950
CIRCUIT BREAKERS					
Excellent	297	11,640	5,014	0	16,951
Fair	1	12,620	11,860	0	24,481
Poor	0	0	1,810	0	1,810
None	0	0	0	7,608	7,608
Total	298	24,260	18,684	7,608	50,850
MOTOR STARTERS					
Excellent	126	2,724	0	0	2,850
Fair	68	4,348	3,435	0	7,851
Poor	0	680	427	70	1,177
None	0	0	0	0	0
Total	194	7,752	3,862	70	11,878
MOTORS					
Excellent	14,650	1,372	1,259	17	17,298
Fair	121	21,930	2,958	0	25,009
Poor	0	0	74	70	144
None	0	0	0	13	13
Total	14,771	23,302	4,291	100	42,464
GENERATORS					
Excellent	104.4	380.7	0	0	485.1
Fair	74.4	279.8	0	0	354.2
Poor	0	0	0	0	0
None	0	0	0	0	0
Total	178.8	660.5	0	0	839.3
DISCONNECT SWITCHES					
Excellent	0	6,287	1,435	0	7,722
Fair	58	426	2,642	0	3,126
Poor	0	402	0	0	402
None	0	0	0	7,365	7,365
Total	58	7,115	4,077	7,365	18,615

(see pp. 681 for the second part of Table 64)

TABLE 64 - POPULATION OF ELECTRICAL EQUIPMENT VERSUS MAINTENANCE QUALITY & NORMAL MAINTENANCE CYCLE

MAINTENANCE QUALITY Card-Type 2 Col. 36	MAINTENANCE, NORMAL CYCLE Card-Type 2 Col. 34 — Less Than 12 Months	12-24 Months	More Than 24 Months	No Preventive Maintenance	Total
	Population: Unit-Years				
SWITCHGEAR BUS - INSULATED**					
Excellent	0	364	12,160	0	12,524
Fair	0	1,706	0	0	1,706
Poor	0	0	0	0	0
None	0	0	0	1,541	1,541
Total	0	2,070	12,160	1,541	15,771
SWITCHGEAR BUS - BARE**					
Excellent	0	1,854	27,580	0	29,434
Fair	0	19,440	2,826	0	22,266
Poor	0	769	0	0	769
None	0	0	0	369	369
Total	0	22,063	30,406	369	52,838
OPEN WIRE (Unit = 1,000 Circuit Feet)					
Excellent	0	2,217	1,014	0	3,231
Fair	0	103	2,630	0	2,733
Poor	0	0	0	0	0
None	0	0	0	680	680
Total	0	2,320	3,644	680	6,644
CABLE (Unit = 1000 Circuit Feet)					
Excellent	600	329	400	0	1,329
Fair	7	7,900	8,519	135	16,561
Poor	0	23	563	35	621
None	0	0	203	9,920	10,123
Total	607	8,252	9,685	10,090	28,634
CABLE JOINTS					
Excellent	0	9,374	311	0	9,685
Fair	12	2,800	23,530	0	26,342
Poor	0	0	1,483	0	1,483
None	0	0	0	12,110	12,110
Total	12	12,174	25,324	12,110	49,620
CABLE TERMINATIONS					
Excellent	2,500	14,290	15,650	0	32,440
Fair	0	1,452	35,200	1,170	37,822
Poor	0	0	845	0	845
None	0	0	0	54,280	54,280
Total	2,500	15,742	51,695	55,450	125,387

**Unit - Number of Connected Circuit Breakers or Instrument Transformer Compartments

failures due to inadequate maintenance, versus 1) failure, months since maintained, and 2) maintenance quality. Failure rate calculations are also given in Tables 67 through 78; these calculations used the population data from Table 64.

Table 79 summarizes the number of failures for all equipment classes combined versus the maintenance quality. Table 80 summarizes the number of failures for all equipment classes combined versus the months since maintained.

GENERAL CONCLUSIONS—MAINTENANCE QUALITY

The maintenance quality is an opinion that was reported by each participant in the survey. The major portion of the electrical equipment population in the survey had a maintenance quality that was classified as excellent or fair. Less than 5 percent of the population in each equipment class (except for motor starters) were classified as poor. Four equipment categories had between 24 percent to 43 percent of the population classified as "none" under maintenance quality; this included cable termination (43 percent), disconnect switches (40 percent), cable (35 percent), and cable joints (24 percent).

Maintenance quality had a significant effect on the percent of all failures that were blamed on inadequate maintenance. In the "poor" category 33 percent of all failures were blamed on inadequate maintenance. This compares with 18 percent for fair maintenance and 12 percent for excellent maintenance. The "none" category for maintenance quality also had 12 percent of all failures blamed on inadequate maintenance; but 82 percent of these failures were for equipment classes that do not require much maintenance (cable, cable terminations, cable joints,

TABLE 65 - PERCENT OF ELECTRICAL EQUIPMENT POPULATION VERSUS MAINTENANCE QUALITY (All Data Taken from Table 64)

MAINTENANCE QUALITY Card-Type 2 Col. 36	TRANSFORMERS	CIRCUIT BREAKERS	MOTOR STARTERS	MOTORS	GENERATORS	DISCONNECT SWITCHES	SWITCHGEAR BUS-INSULATED	SWITCHGEAR BUS-BARE	OPEN WIRE	CABLE	CABLE JOINTS	CABLE TERMINATIONS
	%	%	%	%	%	%	%	%	%	%	%	%
Excellent	54	33	24	41	58	41	79	56	49	5	20	26
Fair	44	48	66	59	42	17	11	42	41	58	53	30
Poor	2	4	10	0+	0	2	0	1	0	2	3	1
None	0+	15	0	0+	0	40	10	1	10	35	24	43
Total	100	100	100	100	100	100	100	100	100	100	100	100

TABLE 66 - PERCENT OF ELECTRICAL EQUIPMENT POPULATION VERSUS NORMAL MAINTENANCE CYCLE (All Data Taken from Table 64)

MAINTENANCE, NORMAL CYCLE Card-Type 2 Col. 34	TRANSFORMERS	CIRCUIT BREAKERS	MOTOR STARTERS	MOTORS	GENERATORS	DISCONNECT SWITCHES	SWITCHGEAR BUS-INSULATED	SWITCHGEAR BUS-BARE	OPEN WIRE	CABLE	CABLE JOINTS	CABLE TERMINATIONS
	%	%	%	%	%	%	%	%	%	%	%	%
Less than 12 Months	1	1	2	35	21	0+	0	0	0	2	0+	2
12-24 Months	58	47	65	55	79	38	13	42	35	29	25	13
More than 24 Months	41	37	32	10	0	22	77	57	55	34	51	41
No Preventive Maintenance	0+	15	1	0+	0	40	10	1	10	35	24	44
Total	100	100	100	100	100	100	100	100	100	100	100	100

TABLE 67 - NUMBER OF TRANSFORMER
FAILURES VERSUS MONTHS SINCE MAINTAINED AND MAINTENANCE QUALITY

MAINTENANCE QUALITY Card-Type 2 Col. 36	FAILURE, MONTHS SINCE MAINTAINED Card-Type 3, Col. 34					Failures per Unit-Year
	Less Than 12 Months Ago	12 - 24 Months Ago	More Than 24 Months Ago	No Preventive Maintenance	Total	
	Number of Failures Due to ALL CAUSES					ALL CAUSES
Excellent	22	11	5	0	38	
Fair	10	26	16	1	53	
Poor	2	1	1	1	5	
None	0	0	0	3	3	
Total	34	38	22	5	99	.00473
	Number of Failures Due to INADEQUATE MAINTENANCE (Card-Type 3 Col. 40)					INADEQUATE MAINTENANCE
Excellent	0	1	2	0	3	.00027*
Fair	1	0	6	0	7	.00075*
Poor	0	0	0	1	1	.00294*
None	0	0	0	0	0	.00000*
Total	1	1	8	1	11	.00053

* Small Sample Size

TABLE 68 - NUMBER OF CIRCUIT BREAKER
FAILURES VERSUS MONTHS SINCE MAINTAINED AND MAINTENANCE QUALITY

MAINTENANCE QUALITY Card-Type 2 Col. 36	FAILURE, MONTHS SINCE MAINTAINED Card-Type 3, Col. 34					Failures per Unit-Year
	Less Than 12 Months Ago	12 -24 Months Ago	More Than 24 Months Ago	No Preventive Maintenance	Total	
	Number of Failures Due to ALL CAUSES					ALL CAUSES
Excellent	13	60	3	1	77	
Fair	18	42	4	1	65	
Poor	0	2	2	0	4	
None	1	0	0	26	27	
Total	32	104	9	28	173	.00340
	Number of Failures Due to INADEQUATE MAINTENANCE (Card-Type 3 Col. 40)					INADEQUATE MAINTENANCE
Excellent	2	1	3	1	7	.00041*
Fair	2	18	2	0	22	.00090
Poor	0	1	2	0	3	.00166*
None	0	0	0	4	4	.00053*
Total	4	20	7	5	36	.00071

* Small Sample Size

TABLE 69 - NUMBER OF MOTOR STARTER
FAILURES VERSUS MONTHS SINCE MAINTAINED AND MAINTENANCE QUALITY

MAINTENANCE QUALITY Card-Type 2 Col. 36	FAILURE, MONTHS SINCE MAINTAINED Card-Type 3 Col. 34					Failures per Unit-Year
	Less Than 12 Months Ago	12 - 24 Months Ago	More Than 24 Months Ago	No Preventive Maintenance	Total	
	Number of Failures Due to ALL CAUSES					ALL CAUSES
Excellent	13	1	4	0	18	
Fair	45	13	8	0	66	
Poor	1	1	2	0	4	
None	0	0	0	0	0	
Total	59	15	14	0	88	.00741
	Number of Failures Due to INADEQUATE MAINTENANCE (Card-Type 3 Col. 40)					INADEQUATE MAINTENANCE
Excellent	1	0	0	0	1	.00035*
Fair	0	1	3	0	4	.00051*
Poor	1	0	1	0	2	.00170*
None	0	0	0	0	0	—
Total	2	1	4	0	7	.00059*

* Small Sample Size

TABLE 70 - NUMBER OF MOTOR
FAILURES VERSUS MONTHS SINCE MAINTAINED AND MAINTENANCE QUALITY

MAINTENANCE QUALITY Card-Type 2 Col. 36	FAILURE, MONTHS SINCE MAINTAINED Card-Type 3 Col. 34					Failures per Unit-Year
	Less Than 12 Months Ago	12 - 24 Months Ago	More Than 24 Months Ago	No Preventive Maintenance	Total	
	Number of Failures Due to ALL CAUSES					ALL CAUSES
Excellent	56	14	7	0	77	
Fair	58	280	90	11	439	
Poor	0	0	2	0	2	
None	0	0	0	0	0	
Total	114	294	99	11	518	.01221
	Number of Failures Due to INADEQUATE MAINTENANCE (Card-Type 3 Col. 40)					INADEQUATE MAINTENANCE
Excellent	8	1	1	0	10	.00058
Fair	2	25	41	2	70	.00280
Poor	0	0	2	0	2	.01390*
None	0	0	0	0	0	.00000*
Total	10	26	44	2	82	.00194

* Small Sample Size

TABLE 71 - NUMBER OF GENERATOR
FAILURES VERSUS MONTHS SINCE MAINTAINED AND MAINTENANCE QUALITY

MAINTENANCE QUALITY Card-Type 2 Col. 36	FAILURE, MONTHS SINCE MAINTAINED Card-Type 3 Col. 34					Failures per Unit-Year
	Less Than 12 Months Ago	12 - 24 Months Ago	More Than 24 Months Ago	No Preventive Maintenance	Total	
	Number of Failures Due to ALL CAUSES					ALL CAUSES
Excellent	14	9	0	0	23	
Fair	1	4	0	0	5	
Poor	0	0	0	0	0	
None	0	0	0	0	0	
Total	15	13	0	0	28	.03360
	Number of Failures Due to INADEQUATE MAINTENANCE (Card-Type 3 Col. 40)					INADEQUATE MAINTENANCE
Excellent	3	0	0	0	3	.00618*
Fair	0	2	0	0	2	.00565*
Poor	0	0	0	0	0	
None	0	0	0	0	0	
Total	3	2	0	0	5	.00596*

* Small Sample Size

TABLE 72 - NUMBER OF DISCONNECT SWITCH
FAILURES VERSUS MONTHS SINCE MAINTAINED AND MAINTENANCE QUALITY

MAINTENANCE QUALITY Card-Type 2 Col. 36	FAILURE, MONTHS SINCE MAINTAINED Card-Type 3 Col. 34					Failures per Unit-Year
	Less Than 12 Months Ago	12 - 24 Months Ago	More Than 24 Months Ago	No Preventive Maintenance	Total	
	Number of Failures Due to ALL CAUSES					ALL CAUSES
Excellent	4	0	1	0	5	
Fair	4	5	4	0	13	
Poor	0	0	16	0	16	
None	0	0	0	67	67	
Total	8	5	21	67	101	.00542
	Number of Failures Due to INADEQUATE MAINTENANCE (Card-Type 3 Col. 40)					INADEQUATE MAINTENANCE
Excellent	0	0	1	0	1	.00013*
Fair	0	4	1	0	5	.00160*
Poor	0	0	0	0	0	.00000*
None	0	0	0	7	7	.00095*
Total	0	4	2	7	13	.00070

* Small Sample Size

TABLE 73 - NUMBER OF SWITCHGEAR BUS-INSULATED FAILURES VERSUS MONTHS SINCE MAINTAINED AND MAINTENANCE QUALITY

MAINTENANCE QUALITY Card-Type 2 Col. 36	FAILURE, MONTHS SINCE MAINTAINED Card-Type 3 Col. 34 — Less Than 12 Months Ago	12 - 24 Months Ago	More Than 24 Months Ago	No Preventive Maintenance	Total	Failures per **Unit-Year
	Number of Failures Due to ALL CAUSES					ALL CAUSES
Excellent	2	3	10	0	15	
Fair	0	4	1	0	5	
Poor	0	0	0	0	0	
None	0	0	0	0	0	
Total	2	7	11	0	20	.00127
	Number of Failures Due to INADEQUATE MAINTENANCE (Card-Type 3 Col. 40)					INADEQUATE MAINTENANCE
Excellent	0	0	6	0	6	.00048*
Fair	0	0	1	0	1	.00059*
Poor	0	0	0	0	0	
None	0	0	0	0	0	.00000*
Total	0	0	7	0	7	.00044*

* Small Sample Size
**Unit = Number of Connected Circuit Breakers or Instrument Transformer Compartments

TABLE 74 - NUMBER OF SWITCHGEAR BUS-BARE FAILURES VERSUS MONTHS SINCE MAINTAINED AND MAINTENANCE QUALITY

MAINTENANCE QUALITY Card-Type 2 Col. 36	FAILURE, MONTHS SINCE MAINTAINED Card-Type 3 Col. 34 — Less Than 12 Months Ago	12 - 24 Months Ago	More Than 24 Months Ago	No Preventive Maintenance	Total	Failures per **Unit-Year
	Number of Failures Due to ALL CAUSES					ALL CAUSES
Excellent	2	1	1	0	4	
Fair	4	6	2	2	14	
Poor	2	0	0	0	2	
None	0	0	0	3	3	
Total	8	7	3	5	23	.00044
	Number of Failures Due to INADEQUATE MAINTENANCE (Card-Type 3 Col. 40)					INADEQUATE MAINTENANCE
Excellent	0	0	0	0	0	.00000*
Fair	1	1	2	0	4	.00018*
Poor	0	0	0	0	0	.00000*
None	0	0	0	1	1	.00271*
Total	1	1	2	1	5	.00009*

* Small Sample Size
**Unit = Number of Connected Circuit Breakers or Instrument Transformer Compartments

TABLE 75 - NUMBER OF OPEN WIRE FAILURES VERSUS MONTHS SINCE MAINTAINED AND MAINTENANCE QUALITY

MAINTENANCE QUALITY Card-Type 2 Col. 36	FAILURE, MONTHS SINCE MAINTAINED Card-Type 3 Col. 34 — Less Than 12 Months Ago	12 - 24 Months Ago	More Than 24 Months Ago	No Preventive Maintenance	Total	Failures per **Unit-Year
	Number of Failures Due to ALL CAUSES					ALL CAUSES
Excellent	0	1	3	0	4	
Fair	1	8	85	0	94	
Poor	0	0	0	0	0	
None	0	0	0	10	10	
Total	1	9	88	10	108	.01628
	Number of Failures Due to INADEQUATE MAINTENANCE (Card-Type 3 Col. 40)					INADEQUATE MAINTENANCE
Excellent	0	1	1	0	2	.00062*
Fair	0	1	30	0	31	.01132
Poor	0	0	0	0	0	— *
None	0	0	0	0	0	.00000*
Total	0	2	31	0	33	.00497

* Small Sample Size
** Unit = 1,000 Circuit Feet

TABLE 76 - NUMBER OF CABLE
FAILURES VERSUS MONTHS SINCE MAINTAINED AND MAINTENANCE QUALITY

MAINTENANCE QUALITY Card-Type 2 Col. 36	FAILURE, MONTHS SINCE MAINTAINED Card-Type 3 Col. 34					Failures per **Unit-Year
	Less Than 12 Months Ago	12 - 24 Months Ago	More Than 24 Months Ago	No Preventive Maintenance	Total	
	Number of Failures Due to ALL CAUSES					ALL CAUSES
Excellent	5	6	2	21	34	
Fair	18	19	16	6	59	
Poor	0	3	2	21	26	
None	0	0	2	95	97	
Total	23	28	22	143	216	.00755
	Number of Failures Due to INADEQUATE MAINTENANCE (Card-Type 3 Col. 40)					INADEQUATE MAINTENANCE
Excellent	0	0	0	0	0	.00000*
Fair	0	2	0	0	2	.00012*
Poor	0	0	2	6	8	.01290
None	0	0	0	12	12	.00119
Total	0	2	2	18	22	.00077

* Small Sample Size
** Unit = 1,000 Circuit Feet

TABLE 77 - NUMBER OF CABLE JOINT
FAILURES VERSUS MONTHS SINCE MAINTAINED AND MAINTENANCE QUALITY

MAINTENANCE QUALITY Card-Type 2 Col. 36	FAILURE, MONTHS SINCE MAINTAINED Card-Type 3 Col. 34					Failures per Unit-Year
	Less Than 12 Months Ago	12 - 24 Months Ago	More Than 24 Months Ago	No Preventive Maintenance	Total	
	Number of Failures Due to ALL CAUSES					ALL CAUSES
Excellent	2	4	0	0	6	
Fair	6	5	1	5	17	
Poor	0	0	0	7	7	
None	0	0	0	15	15	
Total	8	9	1	27	45	.00091
	Number of Failures Due to INADEQUATE MAINTENANCE (Card-Type 3 Col. 40)					INADEQUATE MAINTENANCE
Excellent	0	0	0	0	0	.00000*
Fair	1	0	0	0	1	.00004*
Poor	0	0	0	6	6	.00405*
None	0	0	0	1	1	.00008*
Total	1	0	0	7	8	.00016

* Small Sample Size

TABLE 78 - NUMBER OF CABLE TERMINATION
FAILURES VERSUS MONTHS SINCE MAINTAINED AND MAINTENANCE QUALITY

MAINTENANCE QUALITY Card-Type 2 Col. 36	FAILURE, MONTHS SINCE MAINTAINED Card-Type 3 Col. 34					Failures per Unit-Year
	Less Than 12 Months Ago	12 - 24 Months Ago	More Than 24 Months Ago	No Preventive Maintenance	Total	
	Number of Failures Due to ALL CAUSES					ALL CAUSES
Excellent	3	3	4	0	10	
Fair	3	3	14	3	23	
Poor	0	0	0	1	1	
None	0	0	0	16	16	
Total	6	6	18	20	50	.00040
	Number of Failures Due to INADEQUATE MAINTENANCE (Card-Type 3 Col. 40)					INADEQUATE MAINTENANCE
Excellent	1	1	1	0	3	.00009*
Fair	0	0	5	0	5	.00013*
Poor	0	0	0	0	0	.00000*
None	0	0	0	3	3	.00006*
Total	1	1	6	3	11	.00008

* Small Sample Size

TABLE 79 - NUMBER OF FAILURES VERSUS MAINTENANCE QUALITY FOR ALL EQUIPMENT CLASSES COMBINED

MAINTENANCE QUALITY Card-Type 2 Col. 36	Number of Failures in Tables 67 thru 78		PERCENT of Failures Due to Inadequate Maintenance
	ALL CAUSES	INADEQUATE MAINTENANCE	
Excellent	311	36	11.6%
Fair	853	154	18.1%
Poor	67	22	32.8%
None	238	28	11.8%
Total	1,469	240	16.4%

TABLE 80 - NUMBER OF FAILURES VERSUS MONTHS SINCE MAINTAINED FOR ALL EQUIPMENT CLASSES COMBINED

FAILURE, MONTHS SINCE MAINTAINED Card-Type 3, Col. 34	Number of Failures in Tables 67 thru 78		PERCENT of Failures Due to Inadequate Maintenance
	ALL CAUSES	INADEQUATE MAINTENANCE	
Less than 12 Months Ago	310	23	7.4%
12-24 Months Ago	535	60	11.2%
More Than 24 Months Ago	308	113	36.7%
No Preventive Maintenance	316	44	13.9%
Total	1,469	240	16.4%

and disconnect switches). Thus this 12 percent for "none" is explainable and is not inconsistent with what could be expected.

As maintenance quality decreases from "excellent" to "fair" to "poor," the following equipment classes showed an increasing failure rate from failures blamed on inadequate maintenance: transformers, circuit breakers, motor starters, motors, disconnect switches, switchgear bus—bare, open wire, cable, and cable joints. In some of these cases the sample size is smaller than desirable (less than eight failures) in order to conclusively prove this general statement.

OTHER CONCLUSIONS

Circuit Breakers

Approximately 15 percent of the circuit breaker population had a maintenance quality classified as "none." This compares with less than 1 percent of the population for transformers, motors, and generators.

It is of interest to note that data from Table 60, Part V also show that 15 percent of the circuit breaker population was less than one year old; this compares with less than 3 percent of the population for transformers, motors, and generators. This may possibly account for some of the listings of "none" under maintenance quality reported for failures of circuit breakers.

Motors

Motors with a maintenance quality of "fair" had a failure rate due to inadequate maintenance that was five times higher than motors with excellent maintenance quality.

Open Wire

Open wire with a maintenance quality of "fair" had a failure rate due to inadequate maintenance that was more than ten times higher than open wire with excellent maintenance quality.

DISCUSSION—MAINTENANCE QUALITY

From Table 79 it is possible to calculate for all equipment classes combined the ratio of the number of failures from inadequate maintenance to the number of failures from all other causes. This ratio versus maintenance quality is as follows: poor—0.49, fair—0.22, excellent—

0.13. This is a measure of how much improvement can be obtained by upgrading the maintenance quality from poor to fair to excellent. An excellent maintenance program has only 13 percent more failures added by inadequate maintenance, while a poor maintenance program has 49 percent more failures added by inadequate maintenance.

It is apparent from the data that excellent maintenance quality is very important on open wire and on motors.

It would also appear from the data in Table 65 that essentially everyone in the survey did excellent or fair maintenance on transformers, generators, and switchgear bus—bare. However, on circuit breakers 15 percent of the population had "none" and 4 percent had "poor" on maintenance quality. On motor starters 10 percent had "poor" on maintenance quality. Thus, it would appear that everyone did not maintain circuit breakers and motor starters as well as transformers, generators, and switchgear bus—bare.

One of the drawbacks to the results reported under maintenance quality was that there was no objective definition of "excellent," "fair," or "poor." There are no standards for maintenance quality, and thus this data must be considered to be individual judgment. However, data reported under "failure, months since maintained" does not have this same drawback; this data can be considered factual.

DISCUSSION—FAILURE, MONTHS SINCE MAINTAINED

The data in Table 80 show for all equipment classes combined that there is a close correlation between the percent of failures due to inadequate maintenance and the failure, months since maintained.

Failure, Months Since Maintained	Percent of Failures Due to Inadequate Maintenance
Less than 12 months ago	7.4
12–24 months ago	11.2
More than 24 months ago	36.7

Data from Tables 67 through 78 can also be used to calculate similar correlations for several equipment categories; however, in some cases the sample size is smaller than desirable for adequate statistical confidence.

COMMENTS—NORMAL MAINTENANCE CYCLE

A detailed analysis has not been made of the "maintenance, normal cycle" data in Tables 64 and 66. It is possible that some interesting conclusions could also be drawn from an analysis of this data.

REFERENCES

[1] W. H. Dickinson *et al.*, "Report on reliability survey of industrial plants, part I: Reliability of electrical equipment," *IEEE Trans. Ind. Appl.*, vol. 1A-10, pp. 213–235, Mar./Apr. 1974.
[2] W. H. Dickinson *et al.*, "Report on reliability survey of industrial plants, part II: Cost of power outages, plant restart time, critical service loss duration time, and type of loads lost versus time of power outages," *IEEE Trans. Ind. Appl.*, vol. 1A–10, pp. 236–241, Mar./Apr. 1974.
[3] W. H. Dickinson *et al.*, "Report on reliability survey of industrial plants, part III: Causes and types of failures of electrical equipment, the methods of repair, and the urgency of repair," *IEEE Trans. Ind. Appl.*, vol. 1A-10, pp. 242–249, Mar./Apr. 1974.

Appendix C

Cost of Electrical Interruptions in Commercial Buildings

75 CHO 947-1-1A
pp 123-129

By
Power System Reliability Subcommittee
Industrial & Commercial Power Systems Committee
IEEE Industry Applications Society

P. E. Gannon, *Coordinating Author*
A. D. Patton, *Chairman*

C. E. Becker
M. F. Chamow
W. H. Dickinson
M. D. Harris
S. Wells
C. R. Heising
R. T. Kulvicki
D. W. McWilliams
R. W. Parisian

Industrial and Commercial Power Systems Technical Conference
Institute of Electrical and Electronics Engineers, Inc
Toronto, Canada
May 5-8, 1975

COST OF ELECTRICAL INTERRUPTIONS IN COMMERCIAL BUILDINGS

by

Power Systems Reliability Subcommittee Report
Philip E. Gannon, Coordinating Author1/

Abstract

An IEEE sponsored reliability survey to determine the cost of electrical interruptions in commercial buildings was completed in 1974. The survey form was a simplified version of forms used in 1972 reliability study of industrial plants. The survey included building types and locations, and length and cost of electrical service interruptions. The survey results reflect data from 48 companies covering 55 buildings in the United States. This information is useful in the design of electrical systems for commercial buildings.

Introduction

Knowledge of the cost of power outages, both for normal and critical services, is useful in the design of commercial building power systems, allowing cost-effective judgements to be made with respect to the installation of a second utility company service, an emergency generator, or possibly an uninterruptible power supply.

During 1974, the Reliability Subcommittee of the Industrial and Commercial Power Systems Committee completed a survey of the cost of electrical interruptions in commercial buildings in the United States. Included in this paper are the following results:

1 Cost of power outages to commercial buildings ($ per KWH of undelivered energy).

2 Cost of power outages to commercial buildings ($ per square foot/hr and $ per employee/hr).

3 Critical service loss duration time (length of time before an interruption causes a significant loss).

5 Miscellaneous items relative to provision of auxiliary generators, types of electrical service, and other physical data.

Survey Form

The survey form is shown in Appendix A (two pages). A simple multiple choice or single line fill-in form was utilized in an attempt to reduce the time of the responders, but still provide pertinent data for a meaningful analysis.

Response to Survey

A total of 48 companies reporting on 55 buildings responded to the survey with complete data. Incomplete data, omitting the critical outage cost information was received on 121 additional buildings. Unfortunately, this data was of no value in the present survey. Valid data was submitted almost equally for buildings located in the eastern, central, and western regions of the U.S.A.; with 43 percent of the buildings in downtown areas, 17 percent in urban areas, and 40 percent in suburban areas. Forty-six percent of the buildings were used 5 days per week; 39 percent, 6 days per week; and 15 percent, 7 days per week.

Survey Data Preparation

All of the returned survey forms were reviewed. Useable data was punched onto computer cards for use in data processing.

Survey Results -- Cost of Power Outages

Each respondent was asked to report on the cost of power outages as follows:

1 Dollars per failure -- 15-minute duration, one-hour duration, and greater than one-hour duration; total value of lost operation including wages, damages for delays, loss of computer time, and loss of retail sales minus cost of goods not sold was to be included.

2 Critical service loss duration time -- length of time before an interruption causes a significant loss.

3 Building maximum power demand, and usage, as well as area and number of employees.

The data made it possible to calculate the cost of power outages in terms of dollars per kilowatt-hours of undelivered energy at building peak load.

The average cost of power outages from the survey for the buildings surveyed is given in Table 1.

TABLE 1

AVERAGE COST OF POWER OUTAGES FOR BUILDINGS IN THE UNITED STATES

All commercial buildings	$7.21/KWH not delivered
Office buildings only	$8.86/KWH not delivered

The average maximum demand was 3,095 KW for all commercial buildings reporting outage costs. The maximum demand for the office buildings was only 3,035 KW.

Additional details of the cost of power outages are given in Tables 2, 3, and 4. The tables present additional data including:

1 Outage costs for "office buildings" as a function of duration of outage for three time periods.

2 Effect of computers on outage costs.

3 Relationship of outage costs to: KWH not delivered, to cost per 1,000 square feet per hour of building affected, and to cost per employee per hour affected.

1/ Other members of Sub-Committee: A.D. Patton Chairman; C.R. Heising, Vice Chairman; C.E. Becker; M.F. Chamow; W.H. Dickinson; M.D. Harris; R.T. Kulvicki; D.W. McWilliams; R.W. Parisian; Stanley Wells

TABLE 2

OUTAGE COSTS FOR "OFFICE BUILDINGS"
AS A FUNCTION OF DURATION
(WITH AND WITHOUT COMPUTERS)

	Sample Size	Maximum	Minimum	Average
15-Minute Duration				
Cost/peak KW hr. not delivered	25	$ 22.22	$ 1.50	$ 7.54
Cost/1,000 sq. ft. of bldg./hr.	26	247.6	10.5	63.8
Cost/employee/hr.	26	52.0	3.0	16.0
1-Hour Duration				
Cost/peak KW hr. not delivered	29	$ 24.93	$ 0.64	$ 6.74
Cost/1,000 sq. ft. of bldg./hr.	32	125.00	5.24	53.12
Cost/employee/hr.	32	34.30	1.25	12.22
Duration 1 Hour				
Cost/peak KW hr. not delivered	13	$100.00	$ 0.16	$ 16.16
Cost/1,000 sq. ft. of bldg./hr.	14	320.00	1.05	68.06
Cost/employee/hr.	14	75.80	0.48	16.41

TABLE 3

OUTAGE COSTS FOR "OFFICE BUILDINGS"
AS A FUNCTION OF DURATION
(WITHOUT COMPUTERS)

	Sample Size	Maximum	Minimum	Average
15-Minute Duration				
Cost/peak KW hr. not delivered	11	$ 10.70	$ 1.50	$ 5.84
Cost/1,000 sq. ft. of bldg./hr.	11	107.4	10.54	49.54
Cost/employee/hr.	11	28.56	3.00	12.56
1-Hour Duration				
Cost/peak KW hr. not delivered	13	$ 13.33	$ 0.91	$ 5.30
Cost/1,000 sq. ft. of bldg./hr.	15	120.0	5.24	49.42
Cost/employee/hr.	15	28.57	1.25	10.64
Duration 1 Hour				
Cost/peak KW hr.	3	$100.00	$ 1.97	$ 36.66
Cost/1,000 sq. ft. of bldg./hr.	3	320.00	48.00	156.00
Cost/employee/hr.	3	50.00	4.00	27.52

TABLE 4

OUTAGE COSTS FOR "OFFICE BUILDINGS"
AS A FUNCTION OF DURATION
(WITH COMPUTERS)

	Sample Size	Maximum	Minimum	Average
15-Minute Duration				
Cost/peak KW hr. not not delivered	14	$ 22.22	$ 1.88	$ 8.89
Cost/1,000 sq. ft. of bldg./hr.	15	250.00	16.57	78.21
Cost/employee/hr.	15	52.00	4.00	18.53
1-Hour Duration				
Cost/peak KW hr. not delivered	16	$ 24.93	$ 1.88	$ 8.30
Cost/1,000 sq: ft. of bldg./hr	17	125.00	15.88	54.52
Cost/employee/hr.	17	34.30	4.00	13.62
Duration 1 Hour				
Cost/peak KW hr. not delivered	10	$ 67.66	$ 0.16	$ 9.81
Cost/1,000 sq. ft. of bldg./hr.	11	226.19	1.05	44.08
Cost/employee/hr.	11	75.82	0.48	12.70

TABLE 5

CRITICAL SERVICE LOSS DURATION TIME FOR "ALL BUILDINGS"

	Service Loss Duration Time								
	1 Cycle	2 Cycles	8 Cycles	1 Sec.	3 Sec.	5 Min.	30 Min.	1 Hour	12 Hours
Percent of buildings with critical service loss duration less than or equal to the time indicated.	3%	6%	9%	15%	18%	36%	64%	79%	100%

TABLE 6

CRITICAL SERVICE LOSS DURATION TIME FOR "OFFICE BUILDINGS"

	Service Loss Duration Time								
	1 Cycle	2 Cycles	8 Cycles	1 Sec.	3 Sec.	5 Min.	30 Min.	1 Hour	12 Hours
Percent of buildings with critical service loss duration less than or equal to the time indicated.	5%	10%	15%	25%	30%	50%	70%	75%	100%

TABLE 7

RELATIONSHIP OF AUXILIARY GENERATORS AND SINGLE FEEDER SERVICE TO "ALL BUILDINGS"

	Number of Responses	Buildings with Auxiliary Generation	No Auxiliary Generation and Only Single Feeder
Buildings with computers	23	15	1
Buildings without computers	32	13	7
TOTAL	55	28	8

Survey Results -- Critical Service Loss Duration Time

The amount of time an electrical service can be interrupted before it causes significant losses is a question which our profession has not been able to suitably define. The results of the survey indicate that individual requirements for electrical energy are such that it is probably not possible to establish a general critical service loss duration time. The survey results are shown in Tables 5 and 6.

TABLE 8

TYPE OF ELECTRICAL SERVICE TO "ALL BUILDINGS"

	Number of Responses	Type of Service			
		Single Feeder	Network	Multiple Feeder	Other
Buildings with computers	23	1	8	12	2
Buildings without computers	32	12	10	7	3
TOTAL	55	13	18	19	5

TABLE 9

PHYSICAL DATA -- "ALL BUILDINGS"

Item	Sample Size	Maximum	Minimum	Average
Area, sq. ft. x 10^3	54	2,085	3	400
Number of floors	55	52	1	12
Number of employees	51	7,000	12	1,384
Annual usage - Megawatt hours	52	101,349	210	11,973
Peak Kilowatt demand	52	17,250	95	3,095

TABLE 10

PHYSICAL DATA -- "OFFICE BUILDINGS"

Item	Sample Size	Maximum	Minimum	Average
Area, sq. ft. x 10^3	35	1,600	38	371
Number of floors	35	44	2	13
Number of employees	35	7,000	150	1,651
Annual usage - Megawatt hours	32	51,046	840	9,444
Peak Kilowatt demand	32	17,000	270	3,035

TABLE 11

AVERAGE OF PHYSICAL DATA FOR "ALL BUILDINGS" AND FOR "OFFICE BUILDINGS"

Item	All Buildings	Office Buildings
Megawatt hours/1,000 sq. ft. of buildings area/year	35.5	33.5
Megawatt hours/employee/year	20.2	7.5
Peak Kilowatt demand/1,000 sq. ft. of building area	11.3	11.5
Peak Kilowatt demand/employee	5.0	2.5
Employees/1,000 sq. ft. of building area	3.9	4.7

Thirty-six percent of "all buildings" reporting could be without electrical energy for 5 minutes before the lack of energy was considered to be critical, while 6 percent could be without energy for only 2 cycles and 3 percent for only one cycle before significant losses were incurred.

Fifty percent of the "office buildings" reporting could be without electrical energy for 5 minutes before the lack of energy was considered to be critical, while 10 percent could be without energy for only 2 cycles, and 5 percent for only one cycle before significant losses were incurred.

Precautionary measures taken to minimize critical outages in buildings where computers are installed are indicated in Table 7, where 65 percent (15 of 23) of the buildings reporting have auxiliary generating units. Only 4 percent (1 of 23) of the buildings reporting have no auxiliary generation and are served by a single feeder from the utility company. A like comparison is shown for buildings not having computers; in these instances, 41 percent of the buildings have auxiliary generation and 22 percent are served by single feeders from the utility company.

Table 8 shows the type of electrical service to all buildings reporting. Eighty-seven percent of the buildings with computers have network or multiple feeder service, while 53 percent of the buildings without computers have network or multiple feeder service.

Survey Results -- Demand and Usage Data

Each respondent was asked to report gross floor area, number of floors, number of employees, and electrical energy usage and demand. While not directly related to the subject of this paper, the data is of interest, and will perhaps allow the reader to make a better judgement of the validity of the data presented previously. The details are given in Tables 9, 10, and 11.

It is believed that the employee data for the "All Buildings" category may not be valid, since it appears that not all employees were reported for some multi-function buildings, the office/retail category in particular.

Conclusions and Discussion of Results

1 Cost of Power Outages (Tables 1, 2, 3, and 4)

a There is a wide spread in the cost of power outages (KWH not delivered) in commercial buildings. Even within like types of buildings, with or without computers, there is a great difference in the costs assigned.

b The cost per KWH not delivered increases greatly when the outage duration time exceeds one hour. An exception to this is buildings with computers.

It is probable that for outages of less than one hour, employees may remain partially productive and the temperature of their environment remains tolerable. For longer outages, employees may have to be furloughed for the remainder of the day.

c The cost of power interruptions for buildings with computers varies from $8.89/KWH average for outages of 15-minutes duration to $9.81/KWH for outages of greater than one hour. It is suspected that the small differential is due to the fact that a short duration as well as a long outage renders the computer inoperable, and the employees are either non-productive during this period or repairing possible damage caused by the outage.

d A comparison of the average costs of outages for commercial buildings with that for industrial plants (Reference 1 and 2) is shown in Table 12. The data is interpreted to mean that short-term outages in industrial plants could be more costly than those in commercial buildings, while long-term outages are more costly in commercial buildings.

e Additional information on the cost of power outages in Sweden, Norway, and the United States is contained in Reference 3.

2 Critical Service Loss Duration Time (Tables 5 and 6)

a As would be expected, there is a wide spread in the critical time of a power interruption. This is probably due to the wide variations of type of work being accomplished, the type of equipment involved, and the general work environment. For example, a windowless building in which a sensitive computer operation is performed would be more rapidly affected than a window-wall building performing normal office functions.

b It is suggested that a future survey attempt to define the reasons for the wide variances.

3 Demand and Usage Data (Tables 9, 10, and 11)

a Of the "all building" data reported, the areas averaged 400,000 square feet, 12 floors in height, with an annual usage of almost 12,000 megawatt hours, and a demand of 3,095 KW. Minimum and maximum data were not available.

TABLE 12

COMPARISON OF AVERAGE COSTS OF POWER OUTAGES IN COMMERCIAL BUILDINGS AND INDUSTRIAL PLANTS

Type	Cost
All commercial buildings	$7.21/KWH not delivered
Office buildings	$8.86/KWH not delivered
Industrial plants -- all	$1.89/KW interrupted + $2.68/KWH not delivered

The data for "office buildings" indicate average values within 10 percent of that for "all buildings," except for the number of employees, which is 16 percent greater.

b The average electrical usage for all buildings and for office buildings only is nearly equal when placed on a per unit basis (33.5 KWH/Sq. Ft.) as is the peak demand (11.3 Watts/Sq. Ft. to 11.5 Watts/Sq. Ft.). The relationship of usage and demand to employees does not correlate for all buildings and office buildings only. As mentioned heretofore, the validity of employee data with regard to the Office/Retail category of buildings is questionable. On this basis, no attempt to draw conclusions has been made.

References

1 A.D. Patton, et al, "Report of Reliability Survey of Industrial Plants, Part 4 - Additional Detailed Tabulation of Some Data Previously Reported in the First Three Parts," IEEE I & CPS Conference Record, June 2-6, 1974.

2 W.H. Dickinson, et al, "Report of Reliability Survey of Industrial Plants, Part 2 - Cost of Power Outages, Plant Restart Time, Critical Service Loss Duration Time, and Type of Loads Lost vs. Time of Power Outages," IEEE I & CPS Conference Record, May 14-16, 1973.

3 R.B. Shipley, A.D. Patton, J.S. Denison, "Power Reliability Cost vs. Worth," IEEE Transactions on Power Apparatus and Systems, PAS-91, P. 2204-2212, September/October 1972.

SURVEY FORM ON COST OF ELECTRICAL INTERRUPTIONS IN COMMERCIAL BUILDINGS

INDUSTRY AND GENERAL APPLICATIONS GROUP

RELIABILITY SUBCOMMITTEE OF THE INDUSTRIAL
& COMMERCIAL POWER SYSTEMS COMMITTEE

Electricity is an integral part of our every day life. If it isn't available -- what is its economic effect? Please help us to find out by filling out this form.

Please address reply to:

A. D. Patton
Texas A & M University
Electric Power Institute
College Station, TX 77843

Date ____________

1. COMPANY NAME (Fill in 3-letter abbreviation of name) ____________

2. BUILDING NO. (Fill in sequence number 1, 2, 3, etc. for building(s) reported on) ____________

3. BUILDING TYPE (Check type which best describes your building):

   ☐ Office ☐ Office/Retail Sales ☐ Office/Retail Sales/Apartment

   ☐ Retail Sales ☐ Other (describe) ____________

4. BUILDING LOCATION (Check applicable items):

   ☐ Downtown; ☐ Urban; ☐ Suburban;

   ☐ USA: Eastern; ☐ USA: Central; ☐ USA: Western

5. BUILDING DATA - GENERAL

   Gross Area, square feet ____________

   Number of Floors ____________

   Average Usage of Building: Hours/Day ________ Days/Week ________

   Estimated Number of Office Employees (if any) ____________

   Estimated Annual Retail Sales (if any) ____________

   Is Auxiliary or Emergency Generation Provided: ☐ Yes ☐ No

SURVEY FORM - COMMERCIAL BUILDINGS IN USA

6. BUILDING ELECTRICAL USAGE DATA

Electrical Energy Usage for 12-month Period ____________ KWH

Electrical Maximum Demand for this Period ____________ KW

Type of Service: ☐ Single Feeder; ☐ Network; ☐ Multiple Feeders With Automatic Transfer

☐ Other (Explain) ____________________

7. COST OF A TOTAL INTERRUPTION OF ELECTRICAL SERVICE TO YOUR BUILDING DURING PEAK PERIOD: (Best Opinion - If no interruptions have occurred, assume hypothetical instances)

a) 15-Minute Duration $ ____________

b) 1-Hour Duration $ ____________

c) ____ Hours Duration $ ____________

Does a, b, or c include losses from an "on-line" electronic computer? ☐ Yes ☐ No

For "Office Buildings" loss should include wages of all employees affected, plus any other direct costs incurred including delays, and damage to equipment. This would include any losses from an "on-line" electronic computer.

For "Retail Sales" cost should include estimated loss of sales minus cost of goods not sold, plus cost of any damage incurred.

8. LENGTH OF INTERRUPTION OF ELECTRICAL SERVICE

If there a definitive length of time before an interruption causes a significant loss? ☐ Yes ☐ No

If "Yes", what is maximum time before significant losses will be incurred? ______ Hours ______ Minutes

Appendix D

Reliability of Electric Utility Supplies to Industrial Plants

75 CHO 947-1-1A
pp 131-133

By
Power Systems Reliability Subcommittee
Industrial & Commercial Power Systems Committee
IEEE Industry Applications Society

A. D. Patton, *Chairman*

C. E. Becker
M. F. Chamow
W. H. Dickinson
P. E. Gannon
M. D. Harris
C. R. Heising
R. T. Kulvicki
D. W. McWilliams
R. W. Parisian
S. Wells

Industrial and Commercial Power Systems Technical Conference
Institute of Electrical and Electronics Engineers, Inc
Toronto, Canada
May 5-8, 1975

APPENDIX D RELIABILITY OF ELECTRIC UTILITY SUPPLIES TO INDUSTRIAL PLANTS

RELIABILITY OF ELECTRIC UTILITY SUPPLIES TO INDUSTRIAL PLANTS

by
Power Systems Reliability Subcommittee
Industrial and Commercial Power Systems Committee
A. D. Patton, Coordinating Author[1/]

ABSTRACT

The paper summarizes the results of a 1974 survey of the reliability of electric utility supplies to industrial plants. Results include the average rates of occurrence and durations of power interruptions as a function of type of electric utility supply. This information should help industrial plant operators choose the types of electric utility supplies best suited to their plants.

INTRODUCTION

The electric utility supply reliability survey reported here is a followup to the 1972 survey of the reliability of electrical equipment in industrial plants.[1,2] The 1972 survey showed that the electric utility supply is the most fallible "component" of an industrial plant system and therefore deserves careful consideration.

Certain of the data in the earlier survey were subject to possible error due to misinterpretation of the survey form. Hence, a prime objective of the present survey was to improve the accuracy of data on electric utility supplies. A second objective was to provide more detailed and definitive data on electric utility supply interruption rates and average durations as a function of the number of supply circuits, the type of switching scheme, and the voltage of the supply circuits. A third objective was to obtain data from a larger number of plants than in the 1972 survey thereby permitting interruption rates and average durations to be determined with greater precision. A total of 87 plants provided usable data, almost triple the number of plants providing data on electric utility supplies in the 1972 survey. Survey response broken down by industry is as follows: cement = 2, chemical = 14, metal = 4, petroleum = 30, pulp and paper = 1, rubber and plastics = 4, and other manufacturing = 32.

It should be emphasized that electric utility supply reliability is a function of a number of factors not directly identified in the data presented here. Included in these reliability-influencing factors are line exposure, weather and other environmental conditions, and utility operating and maintenance practice. Thus, the electric utility supply reliability data given in this paper represents average performance and should not be used in preference to specific data when this is available. Methods are available for computing the reliability performance of an electric utility supply when the reliability performance parameters of utility system components are known.[3]

SURVEY QUESTIONNAIRE

The survey questionnaire requested the following data for each electric utility supply.

1. Type of industry
2. Type of electric utility supply
   a. Number of utility circuits supplying the plant
   b. Mode of operation if more than one supply circuit: all circuit breakers normally closed, manual throw-over scheme, or automatic throw-over scheme
   c. Voltage of utility circuits supplying the plant
   d. Type of supply circuits: overhead or underground
   e. A sketch of the electric utility supply system
3. The period of time covered by the survey report. (Respondents were asked to limit their response to the period January 1, 1968 to the present.)
4. The number of interruptions to the plant due to loss of the electric utility supply during the time period of (3).
5. The duration of each electric utility supply interruption, an indication whether service was restored to the plant by a switching operation or by repair or replacement of failed equipment, and, if known, the equipment which failed causing the interruption.

SURVEY DATA SUMMARY AND DISCUSSION

Some respondents to the survey listed voltage dips which caused disruption of plant production as well as complete interruptions of electric utility service. Other respondents commented on production disruptions due to voltage dips without giving details. However, most respondents reported only on complete interruptions of service and this was the intent of the survey. The Subcommittee feels that the sensitivity to voltage dips is a rather unique characteristic of each plant and process and that average interruption rates including voltage dips would not be very meaningful. Therefore, all voltage dip events were removed from the survey data leaving only those interruptions due to complete loss of electric utility service. Hence, the interruption rates given in the summary tables reflect complete loss of electric utility service only. If a plant is sensitive to voltage dips, the rate of such events must be added to the reported interruption rates to obtain the total rate of production disruption due to utility supply troubles.

Almost all respondents indicated that utility supply circuits are overhead rather than underground. Hence, no effort was made to separate supplies with overhead and underground circuits. The data given in the summary tables essentially reflects overhead supply circuits due to the preponderance of such circuits in the survey response.

Preliminary analyses of utility supply interruption rates by industry category indicated no significant differences between industries. Further, there seems to be no good reason why utility supplies of the same type and voltage should differ between industries. Therefore, the data presented in the summary tables is not broken down by industry.

The survey response broken down by number of utility supply circuits, voltage of utility supply circuits, and mode of operation of multiple supply circuit utility supplies is given in Table I.

1/ Members of the Power Systems Reliability Subcommittee are: A. D. Patton, chairman, C. E. Becker, M. F. Chamow, W. H. Dickinson, P. E. Gannon, M. D. Harris, C. R. Heising, R. T. Kulvicki, D. W. McWilliams, R. W. Parisian, and S. Wells.

Table I
Number of Responding Plants
With Electric Utility Supplies
of Various Types

Number of Supply Circuits	
1 circuit	- 20 plants
2 circuits	- 56 plants
3 or more circuits	- 11 plants
Supply Circuit Voltage	
voltage ≤ 15 KV	- 22 plants
15 KV < voltage ≤ 35 KV	- 17 plants
voltage >35 KV	- 48 plants
Switching Scheme of Multiple Circuit Supplies	
all breakers closed	- 45 plants
manual throwover	- ·9 plants
automatic throwover	- 13 plants

Table I shows that two-circuit supplies are the most common among the responding plants. A much smaller number of plants reported three or more supply circuits. All multiple-circuit supplies are combined in the data tables which follow because such supplies are expected to have similar interruption rates and because of the relatively small sample of supplies with three or more circuits. Responses have been broken into three voltage categories corresponding roughly to distribution voltages, subtransmission voltages, and transmission voltages. This was done because electric utility design and operating practice is rather different at these three function levels. Hence, it can be expected that utility supply reliability will be a function of the system level at which service is provided.

Table I indicates that about two-thirds of the responding plants having multiple circuit utility supplies operate with all circuit breakers closed. That is, service is supplied simultaneously over all supply circuits. Service may also be lost, however, by failures in the plant substation or by a widespread failure in the supplying utility's system. Plants having throwover schemes operate with a single circuit providing normal service. Thus, such plants suffer an interruption any time the normal supply circuit fails. The duration of interruption to such plants is usually limited to the time required ro reclose the normal supply circuit or to switch to the alternate supply circuit if the normal circuit is permanently faulted.

Table II summarizes interruption rate and average interruption duration data for single-circuit utility supplies broken down by voltage level. Interruption rates and average durations are given separately for interruptions reported terminated by utility switching operations and by repair or replacement of failed components. Also given are overall interruption rates and average durations.

Tables III and IV show interruption rates and average durations for multiple circuit utility supplies broken down by switching scheme and by voltage level. Table V shows interruption rates and average durations for multiple-circuit utility supplies which operate with all circuit breakers closed broken down by voltage levels. Similar breakdowns by voltage for throwover switching schemes were not possible due to lack of an adequate data base.

Interruption rates and average durations are given in Tables II through V for interruptions where service is restored by: (a) some switching operation or sequence of switching operations in the electric utility system, and (b) repair or replacement of components which failed in the electric utility system. If service can be restored by some automatic or manual switching action in the electric utility system, whether remote or within the utility switchgear at the plant, interruptions are usually much shorter than if repair or replacement of failed components is required to restore service. The reason for providing data on both short-duration switching-terminated interruptions and on long-duration repair-terminated interruptions is because of possible differences in impact on plant operations.

It should be mentioned here that interruption rates and average durations computed from a small number of observed interruptions should be regarded as less accurate than those computed from a larger sample of observations. In particular, Reference [1] shows that interruption rates computed from an observed number of interruptions less than about 8 or 10 may well be in error by plus or minus 50 per cent or more due to random variations alone.

The data of Tables II through V show the expected trends.

(1) Utility supply interruption rates are lowest for multiple circuit supplies which operate with all circuit breakers closed and highest for single-circuit supplies. Tables II and III show that the interruption rate for single-circuit supplies is about six times that of multiple circuit supplies which operate with all circuit breakers closed. Interruption rates for multiple-circuit supplies which operate with a throwover scheme are comparable to those for single-circuit supplies, but throwover schemes have a smaller average interruption duration than single-circuit supplies.

(2) Interruption rates are highest for utility supply circuits operated at distribution voltages and lowest for circuits operated at transmission voltages.

Direct comparisons between interruption rates determined in this survey and in the 1972 survey are not possible in every case, but where possible show somewhat higher values in the present survey. Since the present survey is believed to be more accurate, has a larger data base, and is more up-to-date, the values presented here are to be preferred over those presented in 1972 survey.

REFERENCES

1. Reliability Subcommittee Report, "Report On Reliability Survey of Industrial Plants, Part I: Reliability of Electrical Equipment", IEEE Transactions on Industry Applications, pp. 213-235, March/April 1974.

2. Reliability Subcommittee Report," Report On Reliability Survey of Industrial Plants, Part III: Causes and Types of Failures of Electrical Equipment, the Methods of Repair, and the Urgency of Repair", Ibid., pp. 242-252, March/April 1974.

3. R. Billinton, R. J. Ringlee, and A. J. Wood, Power-System Reliability Calculations, The MIT Press, Cambridge, Mass., 1973.

Table II
Single Circuit Utility Supplies

Voltage Level	Unit-years of History	Number of Interruptions Reported*		Interruptions Per Year**			Average Interruption Duration, Minutes**		
		N_S	N_R	λ_S	λ_R	λ	r_S	r_R	r
v≤15KV	27.62	25	75	.905	2.715	3.621	3.5	165	125
15KV<v≤35KV	12.67	0	21	-	1.657	1.657	-	57	57
v>35KV	71.16	37	60	.527	.843	1.370	1.5	59	37
all	111.45	62	156	.556	1.400	1.956	2.3	110	79

Table III
Multiple Circuit Utility Supplies
All Voltage Levels

Switching Scheme	Unit-Years of History	Number of Interruptions Reported		Interruptions Per Year			Average Interruption Duration, Minutes		
		N_S	N_R	λ_S	λ_R	λ	r_S	r_R	r
all breakers closed	246.17	63	14	.255	.057	.312	8.5	130	31
man. throw-over	42.33	31	5	.732	.118	.850	8.1	84	19
auto. throw-over	64.36	66	11	1.025	.171	1.196	0.6	96	14
all	352.86	160	30	.453	.085	.538	5.2	110	22

Table IV
Multiple Circuit Utility Supplies
All Swiching Schemes

Voltage Level	Unit-Years of History	Number of Interruptions Reported		Interruptions Per Year			Average Interruption Duration, Minutes		
		N_S	N_R	λ_S	λ_R	λ	r_S	r_R	r
$v \leq 15KV$	81.31	52	12	.640	.148	.788	4.7	149	32
$15KV < v \leq 35KV$	78.00	39	5	.500	.064	.564	4.0	115	17
$v > 35KV$	193.55	69	13	.357	.067	.424	6.1	184	34

Table V
Multiple Circuit Utility Supplies
All Circuit Breakers Closed

Voltage Level	Unit-Years History	Number of Interruptions Reported		Interruptions Per Year			Average Interruption Duration, Minutes		
		N_S	N_R	λ_S	λ_R	λ	r_S	r_R	r
$v \leq 15KV$	45.61	8	4	.175	.088	.263	0.7	335	112
$15KV < v \leq 35KV$	52.61	18	1	.342	.019	.361	7.0	120	13
$v > 35KV$	147.95	37	9	.250	.061	.311	11.0	203	49

*N_S and N_R are, respectively, the number of service interruptions terminated by switching and by repair or replacement.

**Interruption rates and average durations subscripted S and R are, respectively, rates and durations of interruptions terminated by switching and by repair or replacement. Unsubscripted rates and duration are overall values.

Appendix E

Report of Switchgear Bus Reliability Survey of Industrial Plants and Commercial Buildings

By
Power Systems Reliability Subcommittee
Power Systems Support Committee
Industrial Power Systems Department
IEEE Industry Applications Society

P. O. O'Donnell, *Coordinating Author*

J. W. Aquilino
C. E. Becker
W. H. Dickinson
B. Douglas
I. Harley
C. R. Heising
D. Kilpatrick

P. E. Gannon, *Chairman*

D. W. McWilliams
R. N. Parisian
A. D. Patton
C. Singh
W. L. Stebbins
H. T. Wayne
S. J. Wells

Industrial and Commercial Power Systems Technical Conference
Institute of Electrical and Electronic Engineers, Inc
Cincinnati, Ohio
June 5-8-1978

Published by
IEEE Transactions on Industry Applications
Mar/Apr 1979 pp 141-147

Report of Switchgear Bus Reliability Survey of Industrial Plants and Commercial Buildings

Power Systems Reliability Subcommittee
Power Systems Support Committee
Industrial Power Systems Department

PAT O'DONNELL, MEMBER, IEEE
COORDINATING AUTHOR[1]

Abstract—The Power Systems Reliability Subcommittee of the IEEE Industry Applications Society has been conducting surveys of the reliability of electrical equipment in industrial plants and commercial buildings. Switchgear bus was included in a previous survey published in 1973 and 1974 [1] and generated some controversy concerning bare and insulated bus. For this reason, and also for an ongoing effect to continually update the 1973 and 1974 survey [1], switchgear bus reliability has been investigated in a new survey in 1977, and the results are presented. Reference is made to a paper [2] given at the 1977 Industrial and Commercial Power Systems Technical Conference on reasons for conducting the new survey.

INTRODUCTION

CURRENT reliability data on failure rate of electrical equipment can provide a valuable tool for the power systems designer or planner. These data can also be a valuable tool for the manufacturer of the equipment concerned.

Many parameters were included in this new survey in an effort to uncover the most influencing factors on the reliability of bare bus and insulated bus and to allow any new obvious and significant applications considerations to be identified. The questionnaire submitted was condensed to a practical and useful form to obtain optimum response in as short of time period as possible.

Results of the survey are presented in tabular form, and discussion is included primarily where adequate response and population data were obtained. Many questions and uncertainties still exist, and the intent of the following presentation is to report the results of the survey with some discussion, but drawing of definite conclusions is left to the reader.

SURVEY FORM

The questionnaire form (Fig. 1) and cover letter used in the survey are included in the Appendix. Total populations data categorize information into major areas of application. An area of primary concern is maintenance because of its obvious relation to failure rate. However, this is the most difficult datum to obtain in complete and uniform format for meaningful results. Responses in this survey did not permit these results to appear, partly due to the respondents' failure to submit information and partly due to the survey format.

Failed unit data were requested in the form shown in the second portion of the questionnaire. The major categories are causes of failure, types of failure, duration of failure, and failed components. This form is less extensive, but more specifically oriented for switchgear bus than in 1973 and 1974 survey [1].

SURVEY RESPONSE

Table I summarizes the survey response including number of buses, companies, and plants. In this survey, bus "unit-year" is defined as the product of the total number of switchgear connected circuit breakers and connected switches reported in a category times the total exposure time. In the previous survey, the unit-year did not include the number of connected switches; that is, only the connected circuit breakers were counted. Table II shows the 1973 and 1974 [1] survey and is included for comparison of responses. The total number of plants in the new survey response is considerably greater than in the 1973 and 1974 survey, but it is interesting to note that unit-year sample size is slightly less. Also some discrepancy appears in the total number of failures reported in Table I and those of some subcategories in tables to follow. This is due to all companies not responding to every category.

SURVEY RESULTS

Insulated and Bare Bus

A major controversy emerged in the results of the 1973 and 1974 survey [1] concerning bare and insulated switchgear bus. Insulated bus, 601-15 000 V, showed a higher failure rate than bare bus, above 600 V, but data were heavily influenced by the chemical industry. The new survey shows the opposite of this, as seen in Table I, with less chemical industry influence. Bare bus, above 600 V, shows a relatively high failure rate, but the sample size is not large, thus making this observation somewhat questionable. With more companies responding in the

Paper ISPD approved by the Power Systems Protection Committee of the IEEE Industry Applications Society for presentation at the 1978 Industrial and Commercial Power Systems Technical Conference, Cincinnati, OH, June 5-9. Manuscript released for publication October 25, 1978.

The author is with El Paso Natural Gas Company, El Paso, TX 79978.

[1] Other members of the subcommittee are Phillip E. Gannon (Chairman), J. W. Aquilino, Carl E. Becker, W. H. Dickinson, Bruce Douglas, Ian Harley, C. R. Heising, Don Kilpatrick, D. W. McWilliams, R. N. Parisian, A. D. Patton, Dr. Chanan Singh, Wayne L. Stebbins, Harold T. Wayne, and Stanley J. Wells.

Company Name and Plant: ______________________

Industry Type: ______________________

Period Reported - From: Month __________ Year __________

To: Month __________ Year __________

Plant Climate: Temperature ______ Relative Humidity ______

Contamination Level and Type: ______________________

Total Population:

Bus No.	No. CB's & SW's	Age of Bus (YRS)	Bus Type and Rating: Bare	Insulated	Outdoor	Indoor	Copper	Aluminum	L-L Voltage (KV)	Current (KA)	System Application: L-L Voltage (KV)	Ungrounded	Solid Ground	Imped. Ground	Maint Cycle (MO)	Maintenance Data: Extent of Maintenance
1																
2																
3																
4																
5																
6																

Failed Unit Data:

Bus No.	Failure Primary Cause	Failure Contributing Cause	Type of Failure: Short L-G	Short L-L	Open	Other	Last Maint. (MO)	Restore Data: Round Clock	Normal Hours	Schedule Later	Failure Duration (Hrs.)	Failed Component and Material

Fig. 1. Switchgear bus reliability survey for metalclad and metal enclosed switchgear bus.

new survey but with less overall unit-year sample size, the failure rate for all bus shows to be slightly higher than in the previous survey. But on breaking this down further, bare bus failure rate is higher and insulated bus failure rate is lower in the new survey.

Table I shows the chemical industry data broken out since it is believed to be a major contributor in the controversy of the 1973 and 1974 survey [1]. In the new survey the chemical industry dominated the number of failures in each category, but did not dominate sample sizes. This supports the argument of some that bus utilized in the chemical industry should have a relatively high failure rate, especially in the use of bare bus.

Table I also shows median outage duration time after a failure of each category, in hours per failure. It is important to emphasize that these data are based on many factors, and without sufficient supplement from respondents concerning operating procedures, maintenance type, spare parts inventory, etc., the data relate to a very general or all-inclusive type of information.

Grounding Type

Survey results are shown in Tables III–V. Inadequate response and the general nature of the questionnaire format prohibit sufficient results for this category. It is believed that grounding type related to failures is important data, but data should be specific, for example, in types of failures in ungrounded systems and in impedance value of impedance grounded systems. This category may be pursued in greater detail in the next survey.

TABLE I
SWITCHGEAR BUS: INDOOR AND OUTDOOR

NUMBER OF COMPANIES	NUMBER OF PLANTS IN SAMPLE-SIZE	NUMBER OF BUSES	SAMPLE SIZE UNIT-YR	NUMBER OF FAILURES REPORTED	INDUSTRY	EQUIPMENT SUB-CLASS	FAILURE RATE FAILURE PER UNIT-YEAR	MEDIAN HOURS DOWNTIME PER FAILURE
39	56	444	51391	54	ALL	ALL	.001050	28
28	36	245	24855	28	ALL	INSULATED ABOVE 600V	.001129	28
25	35	199	26592	26	ALL	BARE (ALL VOLTAGES)	.000977	28
17	23	132	22420	18	ALL	BARE 0-600V	.000802	27
14	18	67	4172	8	ALL	BARE ABOVE 600V	.001917	36
14	19	92	7425	15	PETROLEUM CHEMICAL	INSULATED ABOVE 600V	.002020	40
11	13	135	7002	18	PETROLEUM CHEMICAL	BARE (ALL VOLTAGES)	.002570	28
10	11	83	4707	13	PETROLEUM CHEMICAL	BARE 0-600V	.002761	22
7	8	52	2295	5*	PETROLEUM CHEMICAL	BARE ABOVE 600V	*	48

* Small sample-size.

TABLE II
RESULTS OF PREVIOUS SURVEY PUBLISHED IN 1973 AND 1974 [1]
SWITCHGEAR BUS: INDOOR AND OUTDOOR

NUMBER OF PLANTS SAMPLE-SIZE	SAMPLE SIZE (UNIT-YEAR)	NUMBER OF FAILURES REPORTED	INDUSTRY	EQUIPMENT SUB-CLASS	FAILURE RATE FAILURES PER UNIT-YEAR	ACTUAL HOURS DOWNTIME/FAILURE INDUSTRY AVERAGE	MINIMUM PLT. AVG.	MEDIAN PLT. AVG.	MAXIMUM PLT. AVG.
12	11740	20	ALL	INSULATED 601-15000V	0.001700	261	5	26.8	1613
12	32280	11	ALL	BARE 0-600V	0.000340	550	2	24	2520
5	20560	13	ALL	BARE >600V	0.000630	17.3	6.9	13	48
5	4003	15	PETROLEUM CHEMICAL	INSULATED 601-15000V	0.003750	340	18	26.8	1613
3	17270	10	PETROLEUM CHEMICAL	BARE >600V	0.000580	19.3	6.9	42	48

TABLE III
TYPE OF GROUNDING OVERALL, BUS INSULATED AND BUS BARE

	UNGROUNDED	SOLID-GROUND	IMPEDANCE-GROUND	NOT REPORTED	TOTAL
(Unit-Year) Sample-Size	20262	9787	17280	4062	51391
# FAILURE	17	12	23	2*	54
FAILURE RATE	.000839	.001226	.001331	-	.001050

* Small sample size.

TABLE IV
BUS INSULATED

	UNGROUNDED	SOLID-GROUND	IMPEDANCE-GROUND	NOT REPORTED	TOTAL
(Unit-Year) Sample-Size	4626	4274	14270	1685	24855
# FAILURE	7*	4*	16	1*	28
FAILURE RATE			.001121	-	.001126

* Small sample size.

TABLE V
BUS BARE

	UNGROUNDED	SOLID-GROUND	IMPEDANCE-GROUND	NOT REPORTED	TOTAL
(Unit-Year) Sample-Size	15636	5513	3010	2377	26536
# FAILURE	10	8	7*	1*	26
FAILURE RATE	.000640	.001451			.000980

* Small sample size.

TABLE VI
AVERAGE AGE OF SWITCHGEAR BUS

	ALL	INSULATED	BARE
AGE 1-10 yrs.	6526 unit-year	1899 unit-year	4627 unit-year
>10 yrs.	44596 unit-year	22887 unit-year	21709 unit-year

Age of Bus

Tables VI-VIII illustrate how failures of insulated and bare bus relate to age in this survey. An interesting observation here is that newer bus appears to experience a higher failure rate than older bus. This might be expected if one considers improper installation, new components failure rate, type of construction of new switchgear, etc. As discussed below under "causes" of failures, the logicality of this observation is not consistent.

As incoming data were analyzed, it became apparent that the period reported (it was assumed that the period reported was the period of best kept records) and the age of bus did not correlate as well as expected in every case, a fallacy in the questionnaire format perhaps. Note that the older bus sample size is much larger.

Indoor and Outdoor Bus

The results of this category are summarized in Tables IX-XI below. Table XI shows an overall result of outdoor bus failure rate versus indoor bus failure rate. Outdoor bus shows a higher failure rate than indoor bus, an observation not too surprising.

Failure Duration

Failure duration results are reported in Tables XII and XIII below and categorized into repair on a round-the-clock emergency basis and repair on a normal working hour basis. This adds more meaning to the data in Table I, but would be more meaningful if repair methods were known. Urgency of repair as shown in Table XIV reveals that most repairs were made on an emergency basis. The data of these tables compare very favorable with those of the previous survey.

Type of Maintenance

Response was disappointingly low in this category and results are presented in Tables XV and XVI. The tables show results of maintenance cycles and time since last maintenance in three groups: 1) less than 12 months, 2) 12-24 months, and 3) more than 24 months. This is a very important category regarding reliability, and hopefully the next survey will produce better results.

Causes of Failures

Primary and contributing causes of failures are summarized in Tables XVII and XVIII. As might be expected inadequate maintenance is a large contributor to failures. This does not necessarily follow from the observation above on age of bus. However, defective components are a large primary cause of failures, which is logical for new installations. Correlation between the two tables below is clearly evident from the contributing cause of exposure to contaminants and the primary cause of inadequate maintenance. Exposure to contaminants, which includes dust, moisture, and chemicals, also supports the data showing outside bus with a relatively high failure rate. Inadequate maintenance was reported as the single largest primary cause of failures in the 1973 and 1974 survey [1]. This prompted the effort to survey type of maintenance in the new survey.

TABLE VII
NUMBER OF FAILURES VERSUS AGE

	ALL	INSULATED	BARE
AGE 1-10 yrs.	15	5*	10
>10 yrs.	37	23	14

* Small sample size.

TABLE VIII
FAILURE RATE (FAILURE PER UNIT-YEAR)

	ALL	INSULATED	BARE
AGE 1-10 yrs.	.002298	*	.002161
>10 yrs.	.000829	.001005	.000645

* Small sample size.

TABLE IX
SWITCHGEAR BUS INSULATED

	OUTDOOR	INDOOR
Sample-Size Unit-Year	4275	20356
FAILURE	7*	19
FAILURE RATE	*	.000933

* Small sample size.

TABLE X
SWITCHGEAR BUS BARE

	OUTDOOR	INDOOR
Sample-Size Unit-Year	2750	22339
FAILURE	8	11
FAILURE RATE	.002909	.000492

TABLE XI
SWITCHGEAR BUS (OVERALL)

	OUTDOOR	INDOOR
Sample-Size Unit-Year	7825	42695
FAILURE	15	30
FAILURE RATE	.001917	.000703

TABLE XII
FAILURE DURATION: ROUND CLOCK VERSUS NORMAL HOUR (HOURS DOWNTIME PER FAILURE)

FAILURE REPAIR URGENCY	BUS INSULATED		BUS BARE	
	MEDIAN	AVERAGE	MEDIAN	AVERAGE
ROUND CLOCK	24 hr.	87 hr.	32 hr.	39 hr.
NORMAL HOUR	240 hr.	430 hr.	24 hr.	154 hr.

TABLE XIII
FAILURE DURATION: ROUND CLOCK VERSUS NORMAL HOUR (HOURS DOWNTOWN PER FAILURE)

	BUS INSULATED		BUS BARE	
	ROUND CLOCK	NORMAL HOUR	ROUND CLOCK	NORMAL HOUR
25 PERCENTILE	8 hr.	8 hr.	3 hr.	4 hr.
50 PERCENTILE	24 hr.	240 hr.	32 hr.	24 hr.
75 PERCENTILE	48 hr.	350 hr.	48 hr.	48 hr.

TABLE XIV
FAILURE REPAIR URGENCY

	ROUND CLOCK	NORMAL HOUR	SCHEDULE LATER
BUS INSULATED	64%	28%	8%
BUS BARE	53%	41%	6%

TABLE XV
NUMBER OF SWITCHGEAR BUS-INSULATED FAILURES VERSUS MAINTENANCE CYCLE

	LESS THAN 12 MO.	12-24 MO.	MORE THAN 24 MO.
Sample-Size (Unit-Year)	3563	8812	7253
# FAILURE	2*	13	6*
FAILURE RATE	-	.001475	

* Small sample size.

TABLE XVI
NUMBER OF SWITCHGEAR BUS BARE FAILURES VERSUS MAINTENANCE CYCLE

	LESS THAN 12 MO.	12-24 MO.	MORE THAN 24 MO.
Sample-Size (Unit-Year)	980	10,455	6312
# FAILURE	2*	12	4*
FAILURE RATE	-	.001147	-

* Small sample size.

TABLE XVII
SUSPECTED PRIMARY CAUSE OF FAILURE

BUS INSULATED	BUS BARE	
26%	17%	1. Defective Component
4%	4%	2. Improper Application
7%	9%	3. Improper Handling
7%	13%	4. Improper Installation
19%	22%	5. Inadequate Maintenance
-	18%	6. Improper Operating Procedure
11%	4%	7. Outside Agency - Personnel
26%	-	8. Outside Agency - Other
-	13%	9. Overheating

TABLE XVIII
CONTRIBUTING CAUSE TO FAILURE

BUS INSULATED	BUS BARE	
6.6%	-	1. Thermocycling
3%	8%	2. Mechanical Structure Failure
6.6%	-	3. Mechanical Damage From Foreign Source
-	15%	4. Shorting By Tools or Metal Objects
3%	-	5. Shorting By Snakes, Birds, Rodents, etc.
10%	4%	6. Malfunction of Protective Device
	4%	7. Improper Setting of Protective Device
3%	-	8. Above Normal Ambient Temperature
3%	15%	9. Exposure to Chemical or Solvents
30%	15%	10. Exposure to Moisture
10%	19%	11. Exposure to Dust or Other Contaminants
6.6%	-	12. Exposure to Non-Electrical Fire or Burning
-	8%	13. Obstruction of Ventilation
10%	4%	14. Normal Deterioration from Age
3%	4%	15. Severe Weather Condition
-	4%	16. Testing Error

TABLE XIX
FAILURE TYPE

BUS INSULATED	BUS BARE	
57%	33%	1. Short L-G
40%	60%	2. Short L-L
-	7%	3. Open
3%	-	4. Other

Failure Type

The survey results on types of failures, shown in Table XIX, show a surprisingly high percentage of failures line-to-line.

GENERAL DISCUSSION

At this point it is well to note the confidence intervals of failure rate for bare and insulated bus. Table XX shows the limits for a 90 percent confidence interval. The table illustrates the statistical limits within which 90 percent of the failures could be expected to occur.

Lack of specific details limits the integrity of some data, and as previously indicated not all categories surveyed were reported in this paper, due primarily to small sample sizes and numbers of failures. As with most surveys, accurate data combined with large response are difficult to obtain since response definitely relates to simplicity in questionnaire format. Data of the effect of maintenance on failure rate are highly desirable for obvious reasons, and effort will be made to acquire this data in the future in a meaningful and usable form.

TABLE XX
CONFIDENCE INTERVALS FOR FAILURE RATE λ

FAILURE RATE (λ) FAILURE PER UNIT-YR	INSULATED BUS >600V	BARE BUS > 600V	BARE BUS ≤ 600V
λ L *	.000779	.000958	.000521
λ	.001129	.001917	.000802
λ U *	.001569	.003488	.001203
% DEVIATION - L	31%	50%	35%
% DEVIATION - U	39%	82%	50%

* Upper and lower limits of 90 percent confidence interval for λ.

APPENDIX E SURVEY OF INDUSTRIAL PLANTS AND COMMERCIAL BUILDINGS

APPENDIX

A. D. Patton
Texas A & M University
Department of Electrical Engineering
College Station, Texas 77843

Dear Sir:

RE: Switchgear Bus Reliability Survey for Metalclad and Metal Enclosed Switchgear

The Reliability Subcommittee of the Industrial and Commercial Power Systems Committee requests your cooperation in a survey to determine the reliability of metal-clad and metal-enclosed switchgear bus in industrial plants. The survey is a follow-up to the general reliability survey of plant equipment in 1971 and is intended to provide more meaningful data on switchgear bus. Attached for your information is a report by the subcommittee on reasons for the survey.

The results of the survey will be published in an IEEE paper and are expected to be of value to system planners and designers in the reliability evaluation of alternatives. Individual responses will be held in confidence and only summaries published.

SURVEY INSTRUCTIONS

It is hoped that the survey form is reasonably self-explanatory. Nevertheless, a sample filled-out data sheet is attached for your guidance, and some brief instructions follow. We wish to emphasize that all requested data are important, but it is realized that some of the requested information may be unknown. In such cases, simply provide the information which is known and leave the other spaces blank. We also encourage you to provide explanatory comments on any of your data as you feel appropriate. If additional data sheets are needed, please duplicate the data sheet provided.

General Data

1) It is vitally important that the period reported be given.
2) The plant climate and contamination data should be your general estimates of the requested information.

Total Population Data

1) Using the total population data block, give requested data for all buses *in service during the period reported* whether or not failures have been experienced. (Note the period reported may not exceed the age of a bus. Use separate data sheets for newer busses.)
2) It is vitally important that the number of connected circuit breakers and switches be given for each bus.

Failed Unit Data

1) List each bus failure event separately.
2) Identify the bus in each failure event by specifying the bus number as assigned in the total population data block.
3) Specify failure cause and contributing cause, where known, using the code numbers on the attached sheet.
4) Specify months since bus was last maintained.
5) Check off urgency of restoration effort.
6) Specify time in hours from onset of failure until bus was restored to service.
7) Describe component which first failed, including component material.

Our schedule dictates that responses be received no later than April 1, 1977. Your participation in this project will be greatly appreciated.

Sincerely,

A. D. Patton
Chairman, Reliability Subcommittee

SURVEY QUESTIONNAIRE

Primary Cause of Failure:

1) defective component,
2) improper application,
3) improper handling,
4) improper installation,
5) inadequate maintenance,
6) improper operating procedures,
7) outside agency–personnel,
8) outside agency–other,
9) overheating.

Contributing Cause to Failure:

1) persistent overloading,
2) transient overvoltage,
3) overvoltage,
4) thermocycling,
5) mechanical structural failure,
6) mechanical damage from foreign source,
7) shorting by tools or metal objects,
8) shorting by snakes, birds, rodents, etc.,
9) malfunction of protective device,
10) improper setting of protective device,
11) above normal ambient temperature,
12) below normal ambient temperatures,
13) exposure to chemicals or solvents,
14) exposure to moisture,
15) exposure to dust or other contaminants,
16) exposure to non-electrical fire or burning,
17) obstruction of ventilation,
18) normal deterioration from age,
19) severe weather conditions,
20) loss or deficiency of cooling medium,
21) testing error.

Comments:

REFERENCES

[1] IEEE Committee Report, "Report on reliability survey of industrial plant," *IEEE Trans. Ind. Appl.*, Mar./Apr., July/Aug., and Sept./Oct., 1974. (Part 1–Reliability of electrical equipment; Part 3–Causes and types of failures of electrical equipment, the methods of repair, and the urgency of repair; Part 5–Plant climate, atmosphere and operating schedule, the average age of electrical equipment, percent production lost, and the method of restoring electrical service after a failure; Part 6–Maintenance quality of electrical equipment.)

[2] IEEE Committee Report, "Reasons for conducting a new reliability survey on switchgear bus-insulated and switchgear bus-bare," Industrial and Commercial Power System Tech. Conf., May 1977, Conf. Rec., p. 91–95.